Printed Circuit Board Design with Microcomputers

Printed Circuit Board Design with Microcomputers

TJ Byers

Intertext Publications
McGraw-Hill Book Company

New York St. Louis San Francisco Auckland Bogotá
Hamburg London Madrid Mexico Milan Montreal
New Delhi Panama Paris São Paolo
Singapore Sydney Tokyo Toronto

This book is dedicated to the memory of Henry J. Miller, a caring father

Library of Congress Catalog Card Number 91-72187

10 9 8 7 6 5 4 3 2 1

ISBN 0-07-009558-2

Intertext Publications/Multiscience Press, Inc.
One Lincoln Plaza
New York, NY 10023

McGraw-Hill Book Company
1221 Avenue of the Americas
New York, NY 10020

Composed in Ventura Publisher by High Text Corp.,
Colleyville, Texas.

Preface

Like so many engineering and manufacturing disciplines, the printed circuit board designer is finding the PC an economical alternative to the traditional method of drawing art by hand or using a dedicated CAD workstation. The newest generation of PC circuit board software design packages offer both economy and flexibility, giving the designer maximum benefit at minimal cost. The selection and proper use of these software applications for maximum benefit is what this book is about.

This book specifically addresses the issues involved with designing and manufacturing printed circuit boards using the IBM PC/AT or compatible computers. The book meets the needs of the engineer, technician, and hobbyist alike. It answers the foremost questions of using circuit design and printed circuit board layout on the PC, including the buying decisions and proper use of PC-based circuit design software.

The organization of the book is such that after reading the first two chapters, the reader should be able to turn out a usable circuit board. The book is divided into two parts. The first part looks at the issues of printed circuit board fabrication, from schematic capture to finished board product. The second half takes a serious look at the schematic capture and PCB layout programs on the market today, with emphasis on user needs.

The first chapter is an overview of the printed circuit design process. Basically, it defines the concepts of PC-based circuit board design and how the software accomplishes the various tasks needed to make a finished product.

Chapter 2 describes schematic capture and its nuances. In addition to defining schematic capture jargon, it tells you how to select and use schematic capture software.

Chapters 3 and 4 look at the procedure of design verification using circuit simulation software.

Chapter 5 discusses the process of turning schematic capture netlist files into printed circuit board layouts. The chapter goes into great detail on the selection and proper use of PCB layout software.

Chapter 6 is an introduction to the second half of the book, where we look at 28 popular circuit design programs. The software reviewed includes schematic capture, PCB layout, program interface, and autorouters.

Chapters 7 through 17 contain the actual reviews of the software packages. The chapters are divided by manufacturer, with each chapter containing information on both schematic capture and PCB layout software sold by that manufacturer.

Chapter 18 provides the reader with valuable PC hardware information that is needed to make an intelligent buying decision when looking for a computer for use with PC-based circuit design software. Among the many topics discussed are monitors, expanded memory, and CPU speed.

This is followed by an appendix with exhaustive tables listing all the software's features and cost. The book closes with a glossary of schematic capture and printed circuit board layout terms and an index.

This book covers all the necessary aspects of PC-based printed circuit board design and fabrication. Whether you're new to the field or a veteran, it will help you to become a printed circuit board design expert.

Contents

Chapter

1

PC Board Fabrication

The printed circuit board is the backbone of the electronics industry. Without it many of today's electronic devices would be impossible to make — most notably the personal computer.

Prior to the invention of the printed circuit board in the 1950s, electronic equipment was built completely by hand using mechanical posts for mounting the individual components. The posts were then wired together via insulated copper wires that were often tied in bundles. The process is slow, prone to error, and does not lend itself well to mass production.

The printed circuit overcomes these shortcomings by providing a rigid base for mounting the components and copper tracks for connecting them. Printed circuits come in all shapes and sizes and can be found in nearly all electronic products, ranging from computers to televisions to toys.

Basically the printed circuit is a rigid insulating board (usually of fiberglass) that has a thin layer of copper clad bonded to one or both of its sides. Using a photographic or mechanical template, portions of the copper clad are

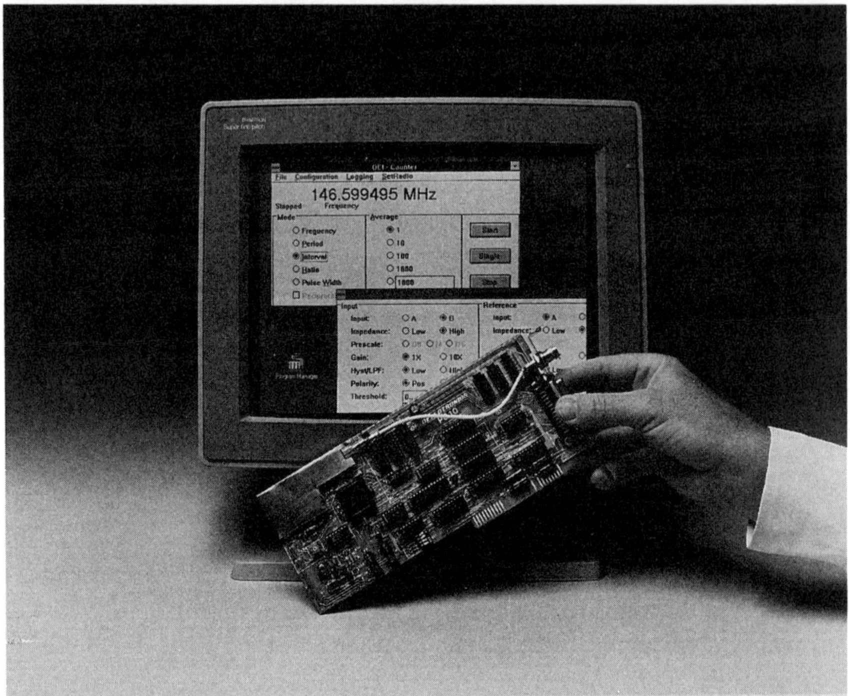

Figure 1-1 Printed circuit boards are the backbone of today's electronic industry.

selectively removed to create a pattern of wires that electrically connects the parts together — hence the name, *printed circuit*. The process is fast, error-free, and well suited for mass production.

Making a Printed Circuit Board

The design and manufacture of a printed circuit is a complex blend of science and witchcraft: witchcraft because it requires insight and second guessing at the onset, and science because it is a linear, rigidly structured process of about a dozen steps. The steps are as follows:

1. Design the circuit
2. Create a schematic
3. Check the schematic for errors
4. Verify the circuit design
5. Define the board size and shape
6. Assign gates to IC packages
7. Place parts on the board
8. Place connecting tracks on the board
9. Check the board for wiring errors
10. Draw the artwork for board fabrication
11. Fabricate the board
12. Test and debug the board

Although the sequence of events is always the same, there is more than one way to perform each step, and the method you choose depends on the type of board you are trying to design, manufacturing constraints, time, and cost. Furthermore, the outcome of a later step may require you to go back to a previous step to make changes to the circuit or the board.

The following is a brief overview of the step-by-step board making process. Even if you are familiar with the printed circuit board design, we urge you to review this section because it introduces terms and concepts which may be new to you, but which are inherent to the automated design process. Subsequent chapters will discuss the steps in greater detail.

Circuit Design

It may seem odd to discuss circuit design as the first step in making a printed circuit, but decisions made here can have far-reaching impact on the worthiness of the final product. Choosing the wrong part type or specifying unrealistic PCB layout tolerances can make fabrication of the printed circuit board too time consuming or too costly, resulting in a product that never gets to market.

There was a time when the design engineer only had to prove that a circuit worked, then pass along the job of making it into a marketable product to someone else. Often the originating engineer never knew if the product ever made it to market — or why it succeeded or failed.

However, in today's fast-paced world; product designers no longer have that luxury. Every day that passes between the concept of a new product and its debut gives the competition an edge in an already crowded marketplace. The solution is to give the circuit designer the tools and knowledge needed to carry the design from conception to finished printed circuit board, thereby minimizing the number of times the printed circuit has to be reworked and reducing the product's time to market. This is where the personal computer gives the engineer an edge, by consolidating the schematic capture and printed circuit board layout in one software package.

Schematic Capture

The second step in the printed circuit design process, *schematic capture,* runs concurrently with the circuit design because it is impossible to evaluate a circuit without a schematic diagram. And what better place to reduce the number of hands a printed circuit design has to go through than at the beginning. Having the circuit designer use schematic capture software rather than scribbling symbols on paper during the circuit design phase eliminates the time-consuming task of redrawing the schematic by a draftsman, resulting in faster turnaround and fewer mistakes.

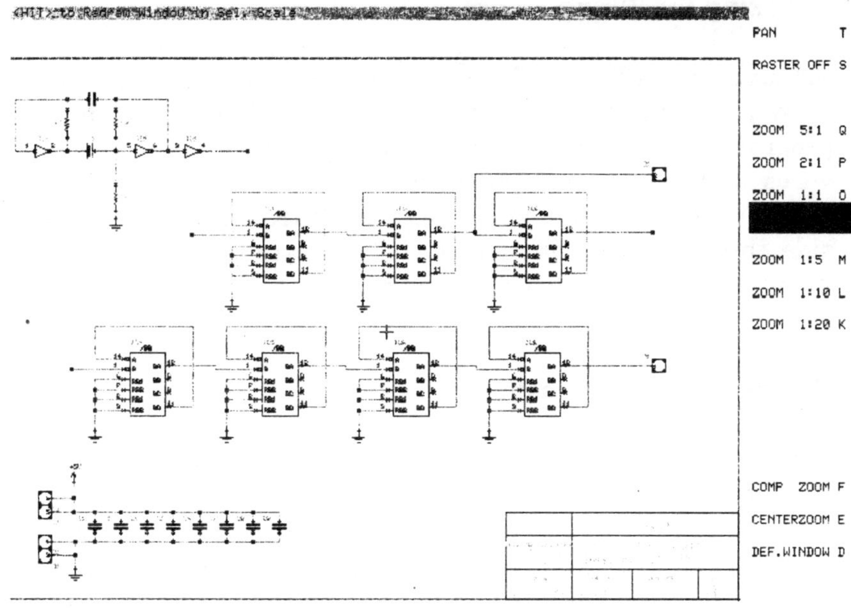

Figure 1-2 Schematic capture software moves the schematic drawing process from the drafting board to the CRT screen.

The schematic is displayed on a CRT screen using standard symbols stored in a component library. The designer selects the desired component from a library menu, then positions the symbol on the screen using a mouse or other input device. Connecting wires between the components are displayed as lines and are also placed on the screen using the mouse.

Changes to the schematic are easily made using the screen editor. Components and connections can be added, deleted, and moved. When moving a part that already has wires attached to it, the editor will drag the connections along with it so that the circuit is not broken. The completed schematic can be drawn on hardcopy using a printer or plotter.

Checking for Wiring Errors

Schematic capture software, however, does much more than just draw pretty pictures. Many schematic capture programs can check for wiring errors, such as meandering wires with no place to go, shorted outputs, floating inputs, and bus contention. Generally, error checking is done outside the drawing screen using a separate software utility included in the schematic capture package.

Circuit Design Verification

Circuit design verification is probably the most important step of the printed circuit design process because it catches empirical design problems before they become hardware problems. It is nearly impossible to predict glitches, phase shift, and skew from a schematic. Problems like this are normally found only after the circuit is built.

But thanks to the power of the PC it is possible to "build" and "test" circuit designs in software instead of hardware, thus eliminating a slow and costly step in the design procedure. The process is called *circuit simulation,* and it is divided into analog simulation and digital simulation.

The information for the circuit simulation comes from the schematic capture program in the form of a *netlist.* The netlist contains a list of all the components in the circuit and their connections to each other. The circuit simulation software reads the netlist, downloads the specified components from the device libraries, and "builds" the circuit in computer RAM. Using complex algorithms, the circuit is analyzed for performance.

The result of the analysis can be displayed in several different formats, but graphics are the most popular. For analog simulation, the graphics show the effects of roll-off,

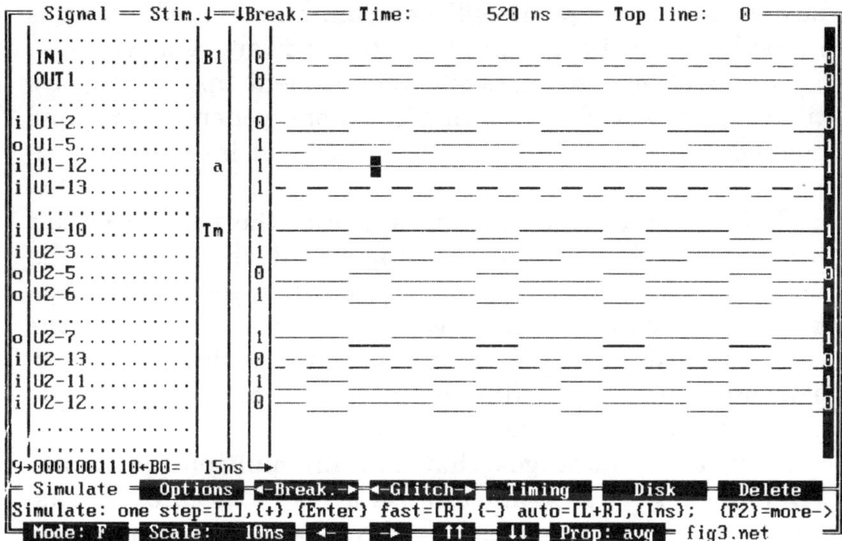

Figure 1-3 Circuit simulation software builds and tests a circuit in computer memory, catching design mistakes before they become hardware mistakes.

phase shift, and spurious oscillation, among other analog parameters. Digital graphics is generally limited to timing diagrams plus glitch and logic violation warnings. Some circuit simulation programs let you make changes to the circuit from the simulation screen, while others have you correct the original schematic capture file and generate a revised netlist for resimulation.

Designing the Printed Circuit Board

Successful completion of the circuit simulation marks the transition between electronic circuit design and printed circuit board design. The amount of thought put into the design up to this point will determine how much effort is needed to turn the design idea into hardware reality.

The next four steps, while independent operations, are tightly interwoven and a change in one generally prompts changes in the other three. They act and interact accordingly.

First, the physical size and shape of the printed circuit board are defined. The resulting board area defines the maximum number of packages that can be mounted on the board, and it must agree with the package count in the design netlist — otherwise, it is back to the drawing board. This step must also define the placement of fixed parts like connectors and switches.

Next, those packages that contain multiple gates are given their gate assignments. As a rule, the gates are assigned to grouped elements within the circuit so that the length of the connecting wire is minimal to avert timing skew problems.

However, gate assignments are very dependent on the next step, which is parts placement. The objective here is to place the parts on the board so that the interconnections form the simplest pattern. But with printed circuit design, the shortest distance between two points is not necessarily a straight line, and you may find that changing gate assignments results in a simpler layout.

The last step in the circuit board layout process is the placement of the interconnecting copper traces using a combination of software autorouting and manual intervention. With very complex designs, it is not unusual to paint yourself into a corner, forcing a change in one or more of the previous steps.

Printed Circuit Board Wiring Verification

After you are convinced that the parts layout and trace pattern are optimal for the design, the board's integrity is

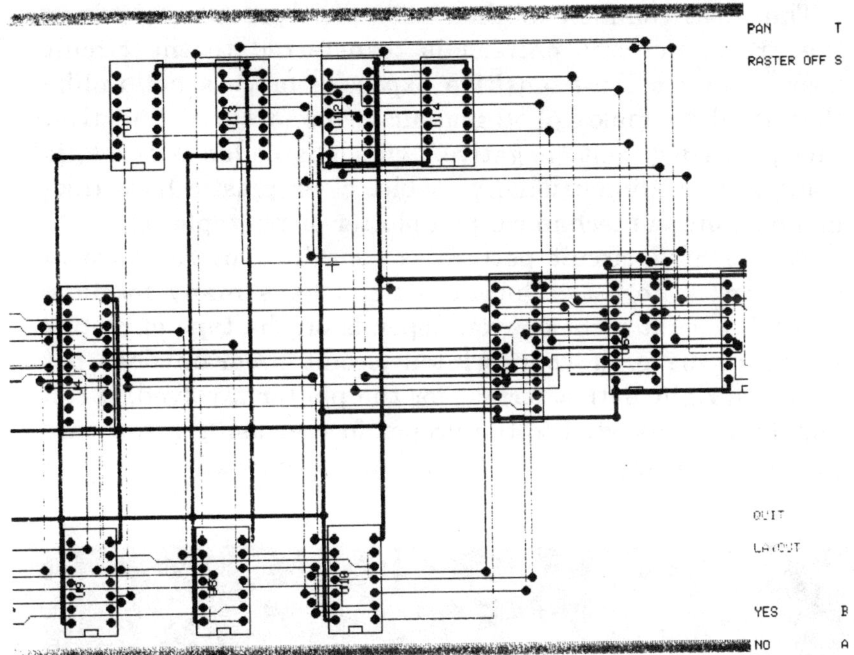

Figure 1-4 PCB layout software replaces the traditional step of pasting decals on clear mylar for printed circuit board layout.

verified. This is done in software using a design rule continuity-checking routine that tests for shorted traces, broken traces, and traces that are too close together to be duplicated on the assembly line.

Preparing the Board Fabrication Artwork

The printed circuit is now ready for fabrication, but first the artwork must be prepared. The type of output required to make the circuit board depends on the manufacturing process involved.

The most common method uses photoprocessing, where the printed circuit pattern is transferred to the circuit board using a light-sensitive exposure process not unlike that used to make photographic prints from a negative. The printed circuit "negative," which is called a *mask,* is printed on a dimensionally stable, clear plastic base (like mylar) using a mechanical pen plotter or photoplotter.

The printed circuit pattern is contained in software in the form of netlists. Output netlists have many formats, and the format right for you depends on the type of plotter it has to drive. If your printed circuit program does not have the right netlist format for the plotter involved, it can usually be converted to the proper format using a software conversion utility.

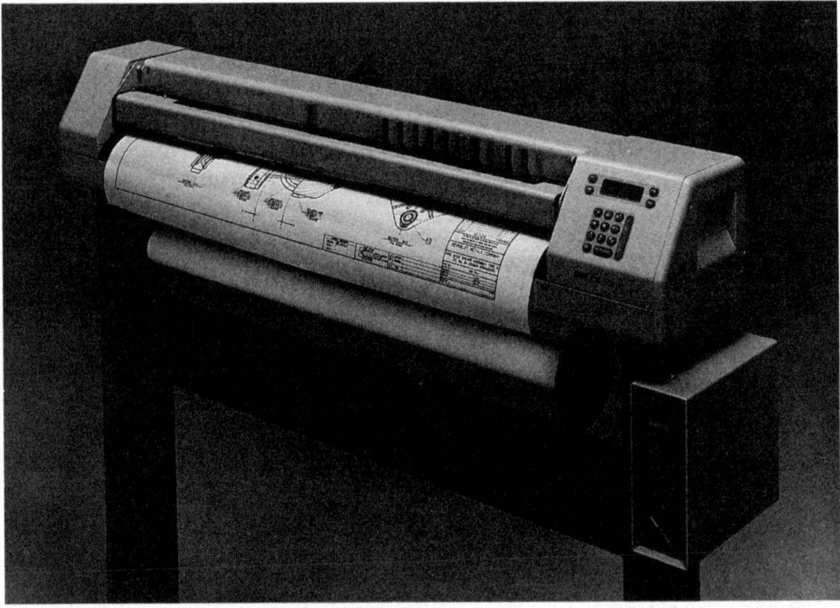

Figure 1-5 Plotters are commonly used to print hardcopy of schematic capture and printed circuit layout designs.

Many electronic equipment manufacturers prefer to have their work done outside by a printed board service. Most of these houses use photoplotters which read netlists that are in the *Gerber* format. Gerber output file format is supported by virtually all PC-based circuit design software. Becoming increasingly popular on the circuit design software scene are Gerber viewers and editors that let you look at a Gerber file and show you exactly how the lines and holes will appear when photoplotted on the circuit board. Gerber editors let you make changes to the file before it is sent to the service house.

Making the Printed Circuit Board

The next step is transforming the artwork into a finished board product. The layer and silkscreen transparencies or Gerber files are first transferred to a photographic negative, which now becomes the only piece of art that will be used for the printed circuit board's fabrication.

The copper-clad board is coated with a photosensitive resist and allowed to dry. The mask is then placed on the sensitized board and exposed to a strong ultraviolet light for a measured length of time. For double-sided boards, both sides have to be exposed before proceeding further. Registration holes in the board and photo masks keep the artwork in alignment during the exposure process.

Exposing the photoresist to ultraviolet light causes it to harden, making it impervious to chemical attack. The board is then etched in a chemical solution, such as ferric chloride, to remove the areas of resist not hardened by the light as a result of the photo mask. The result is a printed circuit board with copper tracks identical to those of the photo mask. The board is then cleaned and made ready for the next fabrication step.

Figure 1-6 Prototype and short production printed circuit boards can be made quickly and inexpensively in-house using outline milling.

An alternative to chemical processing is to mechanically remove the unwanted copper using miniature milling tools, as illustrated in Figure 1-6, using an outlining technique. Rather than laying down narrow conducting tracks, the circuit board is made by cutting narrow insulation channels into standard copper-clad blanks. The insulation channels separate islands of conducting material that now become the tracks. The advantage is that prototype or short production run printed circuit boards can be produced inexpensively within a very short time, usually on the premises. The disadvantage is that not all the copper is removed between the tracks, just enough to form an insulating channel, and the remaining copper can alter the performance of high-frequency circuits.

Holes are then drilled in the printed circuit board to accept component leads. Commonly, this is done by hand using a *drill master*. The drill master is a piece of artwork that clearly shows the location and size of every hole in the printed circuit board. The size of the hole can be identified either by a symbol shape on the drill master, or it may be called out in a separate drill drawing.

Holes can also be drilled by machine using computerized instructions. The instructions used to be fed into the drilling machine via punch tape, but the modern method is to place them in an *Excellon N/C* drill file. The file contains distances from a reference point to the hole and the drill size. Very often the board is plated with a conductive metal such as gold to connect the front and back layers after the holes are drilled.

Board Testing

The final step before assembly is to verify that the board will work, and if not, determine how to fix the problem. Boards are tested in many ways, with visual inspection under a microscope being the most popular. Here the QA (quality assurance) engineer can spot broken tracks, drill hole burrs, unremoved copper, and other items that would prevent the board from working. Electrical testing is also possible, but because of the complexity in setting up a test fixture, it is usually limited to large production boards.

The board is now ready for assembly and soldering. Notice that with perhaps the exception of the last step, board testing, a personal computer was instrumental in the setup and completion of every process. Not only does this allow one person to carry the design and fabrication of a printed circuit board all the way through to completion, it also generates a track record of the entire design process that can be examined to determine reasons for failure and ways to improve a design.

Chapter

2

Schematic Capture

Once the circuit design is finalized, the next step is converting those little black boxes into hardware items like semiconductors and resistors using a *schematic capture* program.

Schematic capture is a critical step because it represents the first step from design concept to hardware, albeit on paper. It is here that you discover that those black boxes need more than imagination to work. Things like power sources and semiconductor types, parameters that are taken for granted in the design phase, suddenly become real and need attention.

However, the traditional method of drawing schematics by hand is both time consuming and prone to error. Fortunately, you can now teach your PC to draw schematics for as little as $99. Schematic capture programs drastically reduce the time it takes to draw a schematic, while at the same time substantially reducing drawing errors.

Furthermore, the netlists that the schematic capture software generates are used by every following design function to arrive at a completed printed circuit board. As such,

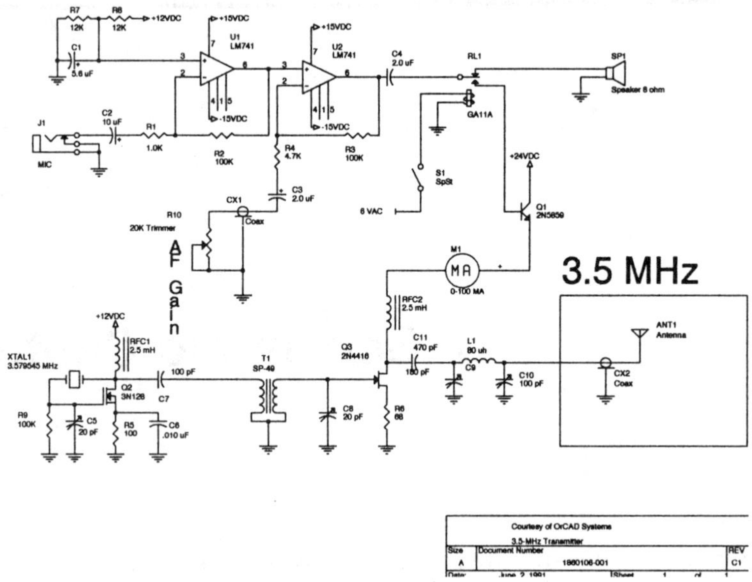

Figure 2-1 Schematic capture programs can produce professional quality schematics.

mistakes made here trickle down to the finished product, making the integrity of the schematic capture phase most important. Fortunately, mistakes made here are also easy to catch and correct, thanks to sophisticated design-checking routines built into many schematic capture packages.

This chapter outlines schematic capture software, explaining how the software works and what features to look for in a program.

CAD in Disguise

Schematic capture programs are basically CAD programs that, over the years, have evolved into special-purpose drawing applications. One of the first PC programs to offer schematic capture support was ***EnerGraphics***, a business graphics charting program with drawing skills. Special library modules were also available for ***AutoCAD*** and ***Generic CADD*** that let you use the CAD program as a schematic capture program. These early beginnings matured into the dedicated CAD-like schematic capture programs we have today.

What sets a schematic capture program apart from a general-purpose drawing program like ***AutoCAD*** is a component library. A component library is nothing more than a collection of symbols that are created using the kernel portion of the CAD program and saved as a block. When a component is needed in a schematic, it can be called from the library rather than having to draw it from scratch.

Schematic capture programs also differ in the way they place and manipulate objects on the screen. Unlike fully featured CAD programs, most schematic capture programs limit their drawing functions to those related to schematic generation. For example, there are no fill, stretch, or component-sizing capabilities. And the one or two line types available are generally predefined, leaving the user no room for creativity.

Schematic capture software has the normal collection of CAD editing tools, like move and delete, but they too are honed for the tasks of editing a schematic. In addition, there are special editing features, like bus management and global replacement of device types, not found in standard CAD programs.

Schematic capture programs also generate hardware-specific lists of schematic items called *netlists*. These netlists are used by circuit simulation and printed circuit board design programs to extend the circuit design process from the schematic screen to a finished printed circuit board. Netlists are extremely important in the making of a printed circuit board and are discussed in more detail later in this chapter.

With but a few exceptions, the hardware requirements for a schematic capture program are the same as for a drawing or a CAD program.

Topping the list of needs is a high-resolution screen display. Screen resolution is measured in pixels, with resolution increasing as the pixel count increases. IBM's EGA and VGA video modes are both excellent choices for use with schematic capture drawings, with the VGA having the edge over EGA in both lower price and higher resolution. Furthermore, many VGA adapter boards also support an 800 × 600 screen that has 35 percent better resolution than VGA. But to use the higher-resolution mode you need a multisync monitor and a special software driver that must be supplied by the schematic capture program.

Screen colors are also important to schematic capture because the program makes extensive use of color if it is available. Components are in one color, connecting wires in another, and nomenclature in yet another. Using colors to identify areas of a circuit makes it easier to choose editing tools when making drawing changes. Generally, the extra dollars spent on a color display is money well spent.

Schematic capture programs also require a lot of hard disk space. You will need anywhere between 1 MB and 10 MB of disk space for the program itself and its libraries, plus space for the schematic files you make (each of which can range from 5K to 500K per schematic, depending on its size).

Because most schematic capture operations are math intensive, the program benefits greatly when a math coprocessor chip is installed in the PC. Typically, a math coprocessor can increase program operations speed by at least two under normal conditions and by as much as tenfold when doing a screen redraw. However, this is not always the case, and if a schematic capture or PCB program does not benefit from a math coprocessor, we tell you so.

Unlike PCB layout hardcopy output, which demands expensive pen plotters, schematic capture programs need nothing more than an inexpensive dot-matrix or laser printer.

Component Library

Central to the schematic capture program is the component library. The component library is a file or group of files that contains the electronic symbols used to describe components on screen and paper.

Generally, the library is made up of smaller, specialized library modules that are linked together. For example, one module would contain nothing but analog ICs, while another might hold the patterns of CMOS devices. Modular libraries are more versatile than a single large library because the architecture allows you to easily expand the capacity of the library by simply adding new modules. Modular architecture is also a great way to save disk space because you only have to load those modules that you normally use and leave the remainder on the floppy disks.

Each schematic capture program has its own way of identifying parts in the component library. The best schematic capture programs list each component by its commercial name, such as 74LS27. The use of commercial names rather than generic ones is also important for the software simulation and printed circuit board layout steps that follow schematic capture. If the schematic capture program simply defines a transistor as NPN silicon, how are the circuit simulation and board layout programs to know whether you are talking about a 2N2222 (a milliwatt small-signal transistor) or a 2N3055 (a 115-watt metal-cased power transistor that requires a heat sink) — both of which are NPN silicon transistors, but neither of which is electrically or physically similar? The use of one or the other makes a big difference in how the circuit performs and how the part is mounted on the circuit board.

Although it is nice to have an exact commercial name for every schematic component, it is not always possible. For example, Schema calls a 15-lead resistor package DRPAC16, a name few would associate with that device.

Fortunately, all programs have a component catalog that shows the library symbols and their names. However, it may take you some time to get a handle on the titles, depending on how obscure the library makes them. The more sophisticated schematic capture programs have a device menu of some kind that lets you highlight the desired component for placement on the schematic.

For components not included in a library, the schematic capture package has a component editor that lets you roll your own. New components can either be created from scratch or by modifying an existing component. Most libraries use bit maps for the component design, which means you have to fill in a matrix dot by dot when building new components. However, there are a few programs with more advanced library editors that let you build components from primitives, like circles and squares. Since the symbols themselves contain the intelligence from which the netlists and parts list are extracted for simulation and board layout, caution needs to be exercised in building new devices. If built improperly, the system may produce schematic diagrams that appear correct but which are electrically or physically incorrect for netlist use.

Drawing Features

Components are called from the library using a display editor. First the editor accesses the library, then calls the component by name. As a rule, display commands are accessed from pull-down menus using a combination of mouse clicks and keyboard strokes. With some programs, the process can be automated using macros (user-defined keystrokes stored in memory) which renders the task of entering a complex routine to a single keystroke.

After a component is selected from the library, the editor places it on the drawing. Common to all schematic capture programs is a grid of dots (visible or invisible) that overlays the drawing page. The grid is used to position the cursor at exactly spaced intervals on the screen. If you try placing the cursor in a position not allowed by the grid, the program moves the cursor to the nearest grid position, a move called grid snap. Grid snap ensures that the component attaches precisely to lines and components already on the schematic with no misalignment. The density of the grid determines the resolution of the drawing. The finer the mesh, the smoother the cursor movement and the more latitude you have over placement of lines and components. Some schematic capture programs allow you to set the size of the grid, while others use a fixed grid. And while most programs let you turn the grid snap off, the software may refuse to recognize two wires as being connected in this mode even though they appear to be properly aligned.

All schematic capture programs have the ability to change the orientation of a component when placing it on the screen. Some programs do it through component rotation, while others store the different orientations in the component library. OrCAD's editor, for example, allows you to rotate a part a full 360 degrees in 90-degree increments, plus generate mirror images of the object in any orientation. Schema, on the other hand, stores the different orientations under separate component names in the component library. Neither method has a particular advantage over the other — unless the orientation you want is not in the component library.

Nearly all schematic capture programs support both ANSI and DeMorgan logic symbols; a few support the new IEEE symbols (see Figure 2-2). DeMorgan logic differs from ANSI logic only in the way their inputs relate to their output logic. ANSI inputs are true when they are positive, whereas DeMorgan inputs are true when their inputs are negative. DeMorgan logic was developed several years ago as an aid in the design of complex logic circuits. Although

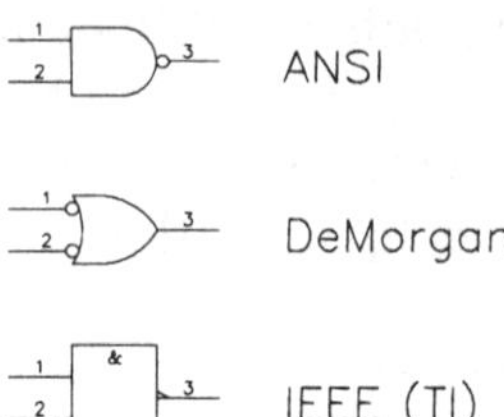

Figure 2-2 Most schematic capture programs support both ANSI and DeMorgan symbols. A few support IEEE symbols.

the function of the device never changes, that is, an ANSI 74LS02 behaves just like a DeMorgan 74LS02, the use of DeMorgan logic often gives the designer insight that may ultimately result in a simpler circuit. Some programs let you toggle between ANSI and DeMorgan from the keyboard, while others store the symbols in the component library under separate names.

After the components are placed on the page, they are interconnected using lines to represent wires. Most schematic programs also support buses (cables that contain several wires) that are drawn as heavier lines. Almost without exception, wires drawn on a schematic run strictly horizontally and vertically; seldom do you find a wire taking a diagonal path across a schematic. In the parlance of CAD, these perpendicular lines are called *ortho* lines (short for orthographic). If you try to take a diagonal path, the drawing editor automatically converts it into an ortho route. However, if the schematic requires a diagonal line or wire, most schematic capture programs let you turn the ortho snap off.

Because schematic capture programs are designed to imitate hand-drawn schematics, several other parameters are permanently fixed. For example, the size of the components are generally predefined, as they would be if drawn on paper using a template. The size of the sheet is also limited

Table 2-1 Standard ANSI and ISO Drawing Sizes

Sheet Size	Dimensions Width	Height
A	$8\frac{1}{2}$ in.	11 in.
B	11 in.	17 in.
C	17 in.	22 in.
D	22 in.	34 in.
E	34 in.	44 in.
A4	218 in.	297 mm
A3	297 in.	420 mm
A2	420 in.	594 mm
A1	594 in.	841 mm
A0	841 in.	1187 mm

to the standard drafting sizes, which range from size A (8½ × 11 inches) to size E (34 × 44 inches). Standard drafting borders and title blocks are also included in the software package, either as a permanent part of the worksheet or as an option. Title blocks can be filled in individually by hand or automatically filled in by the software according to your instructions.

For schematics that are too large for a single sheet, the drawings can be stacked in hierarchical fashion. Some programs also support a linked hierarchy that lets you draw the circuit as a block diagram on a cover sheet. Each block is then drawn in detail using a stack of sheets that defines the circuits within the block. While each block has its own independent stack of drawings, they are linked together so that you can move from one block to another without having to change files. All the sheets are listed under a single file name.

Zoom is provided so that you may zoom in for close-up work or zoom out to get a full view of the worksheet for composition. But the bulk of the work is done using the standard zoom mode. However, standard zoom only shows you a small portion of the entire schematic. To alleviate the problem, most schematic capture programs have autopanning that automatically moves the sheet up, down, or sideways as the cursor touches the borders of the present screen. Some programs use a smooth scrolling autopan that has the sheet in constant motion until you move the cursor away from the screen edge, while others have the screen jump from one area of the worksheet to the adjacent area indicated by the cursor movement — somewhat like flipping through consecutive pages of a book. A few programs require manual panning of the worksheet.

Editing Features

While it is nice to have a software program that takes the drudgery out of drawing schematics, the real advantage of the schematic capture program is its ability to modify the circuit from the screen.

Because hand-drafted schematics take so long to create and tend to accumulate new errors with each revision, they are usually redrawn only when the number of design changes exceeds a certain percentage or after the design is finalized. Schematic capture programs eliminate both problems by allowing you to make the changes on the screen as they become necessary. Several schematic editing tools are used for this task.

Delete and repeat are probably the most used. When using delete, you are generally required to specify and verify the device or line to be deleted to avoid removing something by error. Repeat and delete are often used together so that you can delete a number of lines without having to enter the delete command for each line or device you wish to delete. If you delete something in error, it can usually be replaced using the undo command. Some undo commands

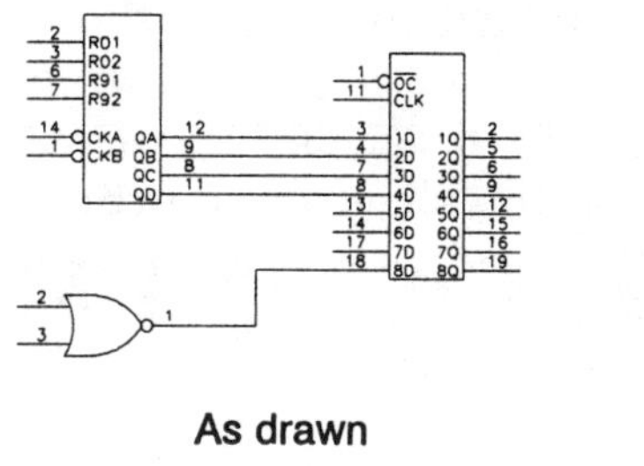

As drawn

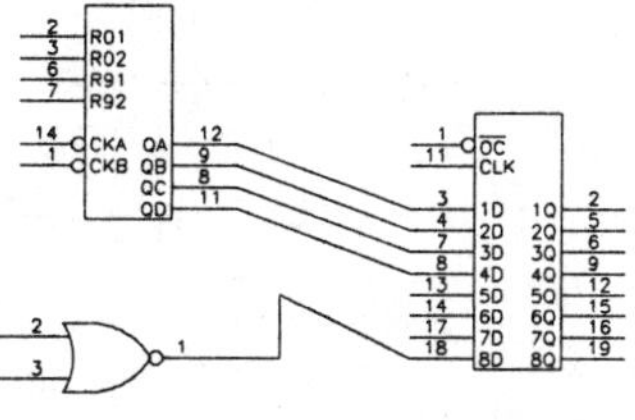

After moving the part
with rubberband enabled

Figure 2-3 The results of moving a part with rubberband enabled.

only restore the last deleted object, while others remember the deletion path you used and undo your changes all the way back to square one.

Moving components is another popular editing recreation. Components are moved for a number of reasons, but normally because their connections to other components change. When a part is moved on the screen, you can retain all its original connections using an editing feature called "rubberbanding." As the name implies, rubberbanding is a technique where the lines connecting two components stretch to accommodate the new position of a device, as illustrated in Figure 2-3.

However, rubberband lines are not ortho, which means you have to reroute them using a cleanup editor that puts the skewed lines back into ortho perspective. Some programs do it automatically, others are manually driven. But either way, you may still have to go into the schematic and manually delete the rubberband lines and replace them with new lines because of the position change. If you wish to move a part without retaining its previous connections, the rubberband feature can be turned off, in which case you end up with a collection of dangling wires that have to be removed manually.

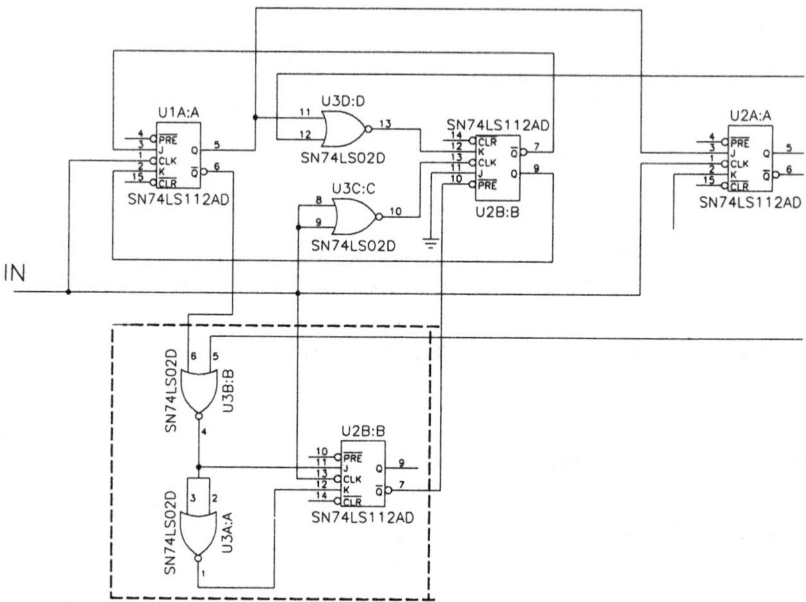

Figure 2-4 The area inside the broken lines is a block and is treated like a single object.

One of the more powerful schematic editing tools is the block editor. A block is a user-defined area on the drawing that the software treats as a single entity, no matter how many elements it may contain, and it is a quick way to deal with massive editing of a worksheet. Blocks may be moved, deleted, or replicated.

Blocks also offer a handy way of creating one-of-a-kind symbols or circuits that are often repeated in a schematic, but which you do not want to save to the component library. For example, if you are drawing a 10-stage frequency divider, instead of drawing 10 identical divider circuits, simply draw the circuit once and save it in a block. You can now move the block to other areas of the worksheet and duplicate the divider circuit as many times as you wish using the copy command. Block circuits are especially useful for drawing large memory arrays.

The find or search command locates parts on a schematic by name and is most helpful when trying to find a part in a complex hierarchical drawing. The find function is often used in conjunction with the replace function to change a component from one type to another, such as changing a 74LS00 logic chip to a 74LS02.

A few schematic capture programs support *macros*. Macros are user-defined keystrokes and/or mouse clicks that are saved to memory so that they may be run again. Some programs have a macro recorder that remembers the sequence as you step through it, while others have you save the sequence in a macro programming language; some let you do both.

Netlists

An electronic system is a collection of electrical components with two or more terminals, called pins, that are interconnected to other pins to form nodes. These nodes, which are called nets in printed circuit board design parlance, document the physical interconnectivity of the various parts in the circuit. A complete set of nets for a circuit is called a netlist.

A netlist is actually composed of two items: the nets themselves and the circuit devices. In the netlist, the relationship between these two items is explicitly documented so that the circuit may be reconstructed by simply reading the netlist. In fact, this is how the printed circuit layout program takes your schematic capture file and makes a circuit board out of it.

Netlist files are generally divided into two sections, the component listings and the pin listings. The relationship

Table 2-2 Typical Schematic Capture Netlists

(PART 1) Listing	Remarks
	1st device
[	Opening bracket
C1	Device reference name
RB.2/.4	Device type (from library)
100uF	Device value
]	Closing bracket
	2nd device
[	Opening bracket
C2	Device reference name
RB.2/.4	Device type (from library)
10uF	Device value
]	Closing bracket
	3rd device
[	Opening bracket
IC1	Device reference name
74LS02	Device type (from library)
14DIP	14-pin DIP package
]	Closing bracket
[	
.	List of devices
.	continues
]	

between the two is made using a pin label. The following discussion describes the format of a typical netlist (see Table 2-2).

The file begins with the component section of the netlist. Each device in the schematic is listed separately, with the information about each component enclosed within a begin-

Table 2-2 (continued) **Typical Schematic Capture Netlists**

(PART 2)

Listing	Remarks
	1st net
(	Opening bracket
GND	Net name
C1-2	Pin 2 of C1
C2-2	Pin 2 of C2
IC1-7	Pin 7 of IC1
)	Closing bracket
	2nd net
(	Opening bracket
C1	Net name
C1-1	Pin 1 of C1
R4-2	Pin 2 of R4
IC1-3	Pin 3 of IC1
)	Closing bracket
(	
.	List of nets
.	continues
)	

The remarks are not part of the Netlists. We included them to describe the function of the Netlists entry.

ning and ending square bracket. The format is shown in the netlist displayed below. The remarks are not part of the netlist; we have used them to show you the meaning of the netlist entry.

The second part of the netlist contains information about the connections within the netlist. Each of the component pins connected to the net is listed. Like the component listing, each net is listed separately. In each case, the component reference name is listed first, followed by a hyphen followed by the pin number of the device connected to that net. To avoid confusion between components and pins, parentheses are used instead of brackets to define the beginning and ending points. The pin section looks as follows; again, the remarks are not part of the netlist, but are included for demonstration purposes. All netlist files are in ASCII format so that they are portable between any PC or workstation.

Printing and Plotting

Another goal of a schematic capture program is to produce a schematic drawing using a printer or a plotter that will be used in conjunction with the PCB layout process somewhere down the line. Two types of drawings are normally generated by the program: draft quality and finished quality.

Draft-quality schematics are used during the drafting of the schematic because they can be quickly generated on a dot-matrix or laser printer. The draft document is then circulated among other engineers and quality control persons to catch errors before they become a permanent part of the design.

However, schematics done on a printer are limited in size, simply because there is no printer made that is big enough to handle an E-size piece of paper. Consequently, the software divides the drawing into pages of 8½ × 11 inches that have to be taped together to get the full picture. Some programs let you fit the schematic to the paper by reducing the size of the drawing. But there is a limit to how much you can reduce an object before it becomes indistinguishable, and a size-E drawing printed on a standard printer paper is not acceptable.

Plotters, on the other hand, are available in sizes from A through E. Moreover, drawings made using a pen plotter are much sharper than those done by a printer. However, plotters are very expensive, with prices in the four- to five-digit range.

Chapter

3

Analog Circuit Simulation Using SPICE

Circuit simulation and verification is the last phase of the circuit design cycle and represents the transition from theoretical engineering to hardware engineering. After this point, the design process changes from symbols on paper to the physical placement of parts and interconnecting traces on a circuit board. Circuit verification is also an extremely important step, because design errors that slip by here become embedded in the hardware and are very costly to correct.

Traditionally, circuit verification consisted of building the circuit on a breadboard and measuring its performance using a battery of laboratory test instruments — a time-consuming and expensive chore. The modern method is to simulate the design on the PC using a software program called a *circuit simulator*. Not only can the circuit be tested in less time and with more accuracy using software circuit simulation, complex testing scenarios such as heat dissipation and worst-case operating conditions are easier to set up and test in software than they are with a breadboard.

Currently there is no one circuit simulation program that does it all, and programs are divided into two types: analog circuit simulation and digital circuit simulation. The two most popular circuit simulation programs for use with printed circuit design on the PC are SPICE and SUSIE. In this chapter, we look at analog circuit simulation using SPICE.

How Circuit Simulators Work

Circuit simulation programs are not new to the computer or the PC. What is new is that circuit simulation programs are now affordable. Simulation programs that sold for $20,000 10 years ago can be obtained today for under $100, making it cheaper and faster to simulate than breadboard. Circuit simulation programs also give us the ability to do "what if" simulations. For example, what if you were to substitute a 0.01 μF capacitor for that 0.015 μF capacitor specified in the design? Using circuit simulation software, you would know the answer in a matter of moments.

Circuit simulation programs simulate the operation of a circuit using mathematical modeling. Basic to the modeling concept are the circuit devices, both active and passive, that are described in mathematical algorithms. For example, resistors are described as current limiting devices with a voltage drop across them using a linear algorithm based on Ohm's law. Transistor models, on the other hand, use a complex algorithm that takes into account forward biasing voltage, reverse current flow, and forward transfer ratio (amplification), among many other operating parameters.

The circuit device algorithms are placed in a file called the *component library*. Each device in the library is identified by its proper name, such as 74LS00 or uA741, and their associated algorithms are completely transparent to the user. The part is simply called by name from the library, and the library supplies the correct algorithm.

Component libraries are generally divided into device types, with one library for analog amplifiers, another for TTL logic chips, and so forth. The simulation software usually comes with two or three popular libraries. Additional libraries can usually be purchased for devices like memory chips, gallium-arsenide (GaAs) parts, programmable logic devices (PLDs), and industrial controllers. Most component libraries also have a device modeler that can be used to describe specialty devices or recently released components not included in the libraries.

Netlists

Like all PC-based engineering programs, circuit simulation software uses netlists to input and output data. The design verification netlist is a listing of all the components in the circuit and their connections to each other. This netlist is normally generated by the schematic capture program but can be created manually.

Circuit simulation begins by loading the netlist into the simulation program. The circuit simulation software then reads the netlist, downloads the specified devices from the component libraries, and "builds" the circuit in computer RAM.

However, netlists are not always interchangeable. The format of the netlist depends on the circuit simulation program used, and not all programs agree on the same format. For instance, some formats require the parts to be listed in an area separate from the connections with special identification names or delimiters, while others let you intermix parts and connections in the netlist with no restriction. It all depends on how the software reads and interprets the netlist.

As a rule, netlists are written in ASCII, which makes it easy to create and modify them using a text editor or word

Table 3-1 Six Elements Required to Make a Working SPICE Netlist

TITLE LINE
ANALYSIS CONTROL STATEMENT
OUTPUT CONTROL STATEMENT
CIRCUIT TOPOLOGY
SIGNAL OR POWER
.END STATEMENT

processor. ASCII files are also portable between different machines, such as IBMs and Apples, which is a big plus if the circuit design has to be transferred to another PC during the design process.

SPICE

SPICE (*S*imulation *P*rogram with *I*ntegrated *C*ircuit *E*mphasis) is far and away the most popular analog circuit simulation program in use today. SPICE was developed by the University of California at Berkeley in the late 1960s and released to the public domain in 1972.

Over the years, SPICE has gone through many upgrades. The most important change came with SPICE2, where the kernel algorithms were upgraded to support advanced integrated system methods, many of which relate to IC performance. SPICE2 has all but replaced SPICE1 as the SPICE of choice and is the predominant version of SPICE in use. SPICE2 has been ported to numerous types of computers and operating systems, including the Macintosh, the IBM PC, and their clones, and is sold under several different brand names by several software companies.

Recently, SPICE2 was upgraded to SPICE3. In the new version, the program was converted from its original Fortran format to C-language for easier portability, and several devices were added to the program's component library, including a varactor, semiconductor resistor, and lossy RC transmission lines. However, the kernel algorithms were not changed, which means that SPICE3 behaves exactly like SPICE2. Furthermore, the new devices found in SPICE3 have been and will continue to be duplicated in SPICE2 using external modeling — but more on that later. So there is no reason to choose SPICE3 over SPICE2 at this time.

Understanding SPICE by Example

The reason for SPICE's popularity lies in its programming simplicity. Unlike many circuit simulator programs, which require special programming tools to create simulation files, all you need to produce a SPICE file is a file editor like DOS's EDLIN or an unformatted word processor that can output an ASCII file. Moreover, because the SPICE files are in ASCII format, files that are created on a Macintosh are identical to those used by the IBM PC or a Sun Workstation, allowing you to exchange SPICE circuit designs on vastly different computers without file modification.

To illustrate how SPICE works, let's put the program through its paces by analyzing the circuit in Figure 3-1. The circuit is an RLC bridge-T bandbass filter often used in audio equalizers. For the sake of argument, let's bias the network at 10 volts and specify a 1-volt ac input signal. The design verification questions are: What is the bandpass frequency, what is the bandwidth, and what are the ac and dc output voltages? Let's see how SPICE solves the problem.

First we draw a schematic of the circuit, then give each component an appropriate label. SPICE nomenclature is fairly consistent with accepted engineering practices, with R representing a resistor, C a capacitor, Q a transistor, and so forth.

Next, the circuit nodes are identified and numbered. A node is any point where two or more wires intersect and connect. There is no order to the numbering, and you may assign node numbers at random and even skip numbers. The exception is ground, which has a reserved value of 0. Just be careful not to assign separate numbers to opposite ends of the same wire, because SPICE will think you are talking about two different nodes and not make the connection.

And that's all there is to it. All you have to do now is list the components by label, node connections, and value — in that order — in an ASCII file with a .CIR extension (e.g., FILTER.CIR). SPICE does the rest. However, SPICE syntax is very inflexible (what programming language isn't?), and unless each component is described explicitly with all the parameters in the right order, the simulation won't work. The completed SPICE file for Figure 3-1 is shown in Table 3-2.

Although we listed the resistors in ascending order, there is no set order to a netlist. Any voltage source, resistor, or other device may be placed anywhere you want in the list, because SPICE first reads the netlist in its entirety, organizes the contents to its liking, then compiles the circuit into a run-time file (in exactly the same way Pascal and C source codes are compiled into programs). The only restrictions are that the file begins with a file name and ends with an .END statement, and that you use uppercase letters in the file.

When describing a SPICE component, you begin with the device name followed by its node attachments and then its value. For example, the entry

```
R1 1 0 200
```

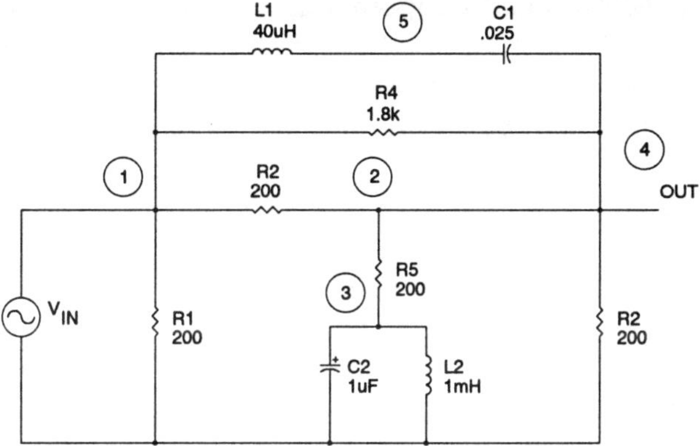

Figure 3-1 Bridge-T bandpass filter.

describes a resistor with a value of 200 ohms connected to nodes 1 and 0. Optional parameters, like temperature coefficient and breakdown voltage, can be appended to the part value for in-depth analysis and worst-case studies.

SPICE netlists can also be generated by most schematic capture programs, such as *Schema* or *OrCAD*. The process is as simple as drawing a schematic using the capture software then requesting a netlist. However, very few schematic capture programs actually produce a *working* SPICE netlist. Almost always, the opening and closing lines are missing. But more important, most schematic capture software lacks device modeling (SPICE descriptions of transistors, etc.), which means you have to go into the netlist and manually insert the missing lines and models. But it beats having to write the whole netlist by hand and lessens the chance for error.

Circuit Analysis and Simulation

After SPICE compiles a run-time version of the file, it uses its kernel algorithms to analyze the circuit, first for dc parameters, then ac performance (when requested). The re-

Table 3-2 SPICE File for Figure 3-1[a]

```
BRIDGE-T BANDPASS FILTER
.AC DEC 20 1KHZ 10KHZ
.PRINT AC V(4) VDB(4)
R1 1 0 200
R2 1 2 200
R3 2 4 200
R4 1 4 1.8K
R5 2 3 200
R6 4 0 200
C1 5 4 .025UF
C2 3 0 1UF
L1 1 5 40MH
L2 3 0 1MH
VIN 1 0 DC 10V AC 1V
.END
```

[a] SPICE cicuit designs are simple ASCII files. There is no order to a SPICE file, and you may place the lines in any order you wish. However, the netlist must open with a file name and close with an .END statement, and all entries must be in uppercase.

sult is a computer simulation of the circuit under the specified operating conditions. The simulated circuit parameters are recorded in ASCII format in a special SPICE output file. The SPICE simulation of our filter network in Figure 3-1 is shown in Table 3-3.

SPICE simulation is broken down into stages, with the first step being an analysis of the netlist to see if there are any errors. It cannot check for all errors, simply because SPICE cannot read your mind, but it does check the syntax of all entries for errors and makes sure that every node has two or more connections. If an error occurs, an error message is inserted below the line causing the error, making netlist debugging almost effortless.

Table 3-3 SPICE Simulation of Filter Network in Figure 3-1

```
**** 2/27/90 ***** IS SPICE  1.41 12/12/87*****11: 9:21*****
 BRIDGE-T BANDPASS FILTER

**** INPUT LISTING                 TEMPERATURE =   27.000 DEG C
 *************************************************************
 .AC DEC 20 1KHZ 10KHZ
 .PRINT AC V(4) VDB(4)
 R1 1 0 200
 R2 1 2 200
 R3 2 4 200
 R4 1 4 1.8K
 R5 2 3 200
 R6 4 0 200
 C1 5 4 .025UF
 C2 3 0 1UF
 L1 1 5 40MH
 L2 3 0 1MH
 VIN 1 0 DC 10V AC 1V
 .END

**** 2/27/90 ***** IS SPICE  1.41 12/12/87*****11: 9:21*****
 BRIDGE-T BANDPASS FILTER

**** SMALL SIGNAL BIAS SOLUTION TEMPERATURE =   27.000 DEG C
 *************************************************************
NODE VOLTAGE     NODE VOLTAGE     NODE VOLTAGE     NODE VOLTAGE
( 1) 10.0000     ( 2)   4.1667    ( 3)    .0000    ( 4)  2.5000
( 5) 10.0000
     VOLTAGE SOURCE CURRENTS
     NAME        CURRENT
     VIN       -8.333D-02
     TOTAL POWER DISSIPATION   8.33D-01  WATTS
**** 2/27/90 ***** IS SPICE  1.41 12/12/87*****11: 9:21*****
 BRIDGE-T BANDPASS FILTER

**** AC ANALYSIS                   TEMPERATURE =   27.000 DEG C
 *************************************************************
     FREQ         V(4)          VDB(4)
  1.00000E+03     2.509E-01   -1.201E+01
  1.12202E+03     2.511E-01   -1.200E+01
  1.25893E+03     2.514E-01   -1.199E+01
```

Table 3-3 (continued) SPICE Simulation of Filter Network in Figure 3-1[a]

```
1.99526E+03      2.544E-01  -1.189E+01
2.23872E+03      2.561E-01  -1.183E+01
2.51189E+03      2.587E-01  -1.174E+01
2.81838E+03      2.629E-01  -1.160E+01
3.16228E+03      2.706E-01  -1.135E+01
3.54813E+03      2.862E-01  -1.087E+01
3.98107E+03      3.254E-01  -9.752E+00
4.46684E+03      4.662E-01  -6.628E+00
5.01187E+03      9.983E-01  -1.499E-02
5.62341E+03      4.882E-01  -6.228E+00
6.30957E+03      3.303E-01  -9.622E+00
7.07946E+03      2.879E-01  -1.081E+01
7.94328E+03      2.714E-01  -1.133E+01
8.91251E+03      2.634E-01  -1.159E+01
1.00000E+04      2.589E-01  -1.174E+01


          JOB CONCLUDED
           TOTAL JOB TIME            14.99
------------------------------------------------------------
```

[a] Like SPICE netlist files, SPICE output files are in ASCII format, making it easy to move the file among different PCs and to print a hardcopy using any type of printer.

If no errors are detected, SPICE does a dc analysis of the circuit, calculating the bias voltage for every node in the circuit and displaying the current drain from the power sources. SPICE also measures the total power dissipation of the circuit. From this analysis, we find that the dc output voltage is 2.5 volts, the first answer to our problem.

Next, SPICE does an ac analysis of the circuit using simulated signal generators. SPICE is quite adept in this area, in that it is able to emulate several different kinds of signal generators, including sine, sweep, and pulse, among others.

For our tests we need a sweep generator, which we describe in the second line of the netlist. The .AC label is a special SPICE command called a *control statement.* (Note that all SPICE control statements begin with a dot or period.) The statement says that we want a sweep generator with a logarithmic sweep of 20 points per decade (DEC 20) beginning at 1 kHz and ending at 10 kHz. We could just as easily have selected a linear (LIN) or an octave (OCT) sweep rate and set the sweep between any two frequencies we wanted, including frequencies from subaudio to the gigahertz range.

Next we need to monitor the output for signal. This is done using the .PRINT control statement shown on line three. Like .AC, .PRINT is a unique control statement that simulates a wide variety of measuring instruments, including several types of voltage and current meters. For our simulation we need both an rms ac voltmeter [V(4)] and a decibel voltmeter [VDB(4)]. The numbers inside the parentheses indicate which node is being measured. As it turns out, our simulation looks at only one node, but we could have specified any number of nodes.

The .PRINT statement generates a table of values for each point on the generator's sweep. The results of our circuit simulation are shown in the AC ANALYSIS section of Table 3-3. Looking at the results, you will notice that the filter peaks at 5 kHz with a voltage output of 750 mV and an attenuation factor of 12 dB, answering two more of the circuit's unknowns. For the final answer, we look at the VDB(4) portion of our table, which shows us that the signal is down 6 dB at 4.5 kHz and 5.5 kHz, giving the filter a bandwidth of 1 kHz. Problem solved.

Had we requested more measurements, SPICE would have continued with the simulation. For example, we could have measured the transient response time, noise factor, distortion, and phase shift of any or all nodes in the circuit over a wide range of operating conditions and temperatures. Such analysis is extremely valuable for seeing how the circuit behaves under extreme environmental conditions, allowing us to optimize or troubleshoot the design.

Device Modeling

SPICE can also analyze circuits that use active devices, like diodes and transistors. However, the process requires that you provide substantially more information about the component. To avoid redundancy in describing active components, SPICE lets you describe the device in detail only once using a method called *device modeling*.

With device modeling, the equations needed to describe the device are built into the SPICE program. The only thing you have to do is fill in the blanks. Once a device is defined, it can be used as many times as you wish in the netlist.

To demonstrate the use of device modeling, let us use the circuit in Figure 3-2 as an example. In this circuit two 2N2222 transistors are cascaded to create a darlington amplifier. Resistors RB1 and RB2 are the biasing resistors, RL is the load resistor, and RE is the emitter resistor; Vin and Vout are the input and output ports, respectively. Examining the circuit, we see that the SPICE netlist contains four resistors, two transistors, and six nodes (including ground). A complete netlist is given in Table 3-4.

Notice that the transistors are treated exactly like a resistor or capacitor, with the node connections listed first and followed by a part value. In this case, the part value refers SPICE to the QN2222 transistor model at the bottom of the netlist. The .MODEL label is a control statement that tells SPICE to plug these values into its transistor equation. For each different type of transistor or other modeling device used in a circuit, you must provide a separate model description.

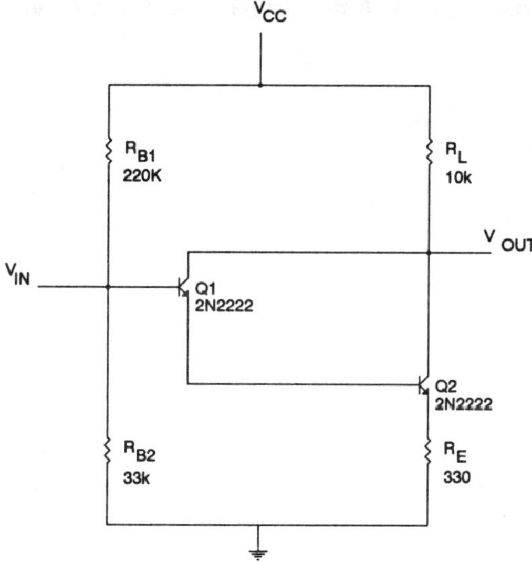

Figure 3-2 Schematic of darlington amplifier.

Subcircuits

When modeling devices more complex than discrete components, SPICE has you describe the device in a *subcircuit.* Subcircuits are used to model everything from ICs to transformers. An example of a SPICE netlist using a subcircuit is shown in Table 3-5.

The circuit described, shown in Figure 3-3, is an active bandpass filter using a uA741 operational amplifier. As before, we begin by drawing the schematic, labeling the components, and assigning numbers to the circuit nodes. Note that the IC is drawn in conventional schematic format as a single element with five external connections.

The actual description of the uA741 IC begins with the line .SUBCKT UA741 and ends with .ENDS. In the uA741 subcircuit netlist you will find the usual assortment of re-

Table 3-4 Example of a SPICE Netlist Using Device Modeling[a]

```
DARLINGTON AMPLIFIER
VCC 2 0 DC 10V
RB1 1 2 220K
RB2 1 0 33K
RL 2 3 10K
RE 5 0 330
Q1 3 1 4 QN2222
Q2 3 4 5 QN2222
.MODEL QN2222 NPN(IS=1.9E-14 BF=150 VAF=100
+ IKF=.175 ISE=5E-11 NE=2.5  BR=7.5 VAR=6.38
+ IKR=.012 ISC=1.9E-13 NC=1.2  RC=.4  XTB=1.5
+ CJE=26PF TF=.5E-9 CJC=11PF  TR=30E-9
+ KF=3.2E-16 AF=1.0)
.END
```

[a] SPICE contains device model equations for diodes, bipolar transistors, junction FETs, and MOSFETs.

sistors and capacitors used to specify input and output impedance characteristics. For example, RL and CL describe the resistive and capacitive factors of the uA741's output pin.

In some subcircuits you may find that individual diodes and transistors are used to describe their inner workings, but more often than not voltage and current sources are used to emulate the collective actions of these components. In our model, for example, VOFST is a voltage source that sets the IC's input offset voltage at one millivolt, and IBN is a current sink that sets the bias current of the inverting input at 100 nanoamps. Furthermore, you will notice that the noninverting input bias current, IBP, is set at 80 nA, giving the op amp an offset bias current of 20 nA, which is typical for this IC.

Table 3-5 Subcircuits Used to Describe Complex Devices, Such as ICs and Transformers

```
OP AMP BANDPASS FILTER
.AC DEC 20 1 100KHZ
.PRINT AC V(5)
X1 3 4 5 7 6 UA741
R1 1 2 20K
R2 0 2 1K
C1 2 3 .01U
C2 2 5 .01U
R4 3 5 100K
R3 0 4 100K
VEE 6 0 -15
VCC 7 0 15
V1 1 0 SIN 0 1 AC 1

.SUBCKT UA741  2    3    6     7      4
*PINOUTS      -IN  +IN  OUT   VCC    VEE
RP 4 7 10K
RXX 4 0 10MEG
IBP 3 0 80NA
RIP 3 0 10MEG
CIP 3 0 1.4PF
IBN 2 0 100NA
RIN 2 0 10MEG
CIN 2 0 1.4PF
VOFST 2 10 1MV
RID 10 3 200K
EA 11 0 10 3 1
R1 11 12 5K
R2 12 13 50K
C1 12 0 13PF
GA 0 14 0 13 2700
C2 13 14 2.7PF
RO 14 0 75
L 14 6 30UHY
RL 14 6 1000
CL 6 0 3PF
.ENDS
.END
```

The node numbering within a subcircuit is independent of the node numbering of the main netlist, and in no way do the two sets of numbers conflict. Duplication of node numbers is not a problem. Furthermore, you can nest subcircuits; that is, you can call another subcircuit into your subcircuit as an element of its design. Like device modeling, you only have to define a subcircuit once, but you must have a separate subcircuit for each different device type. The subcircuit model of the uA741 will not work for an LM3900 Norton amp.

Digital Modeling

SPICE can also perform digital simulations. As a rule, you are better off using a purely digital simulator like SUSIE to measure logical time sequences, but there are times when the two technologies come together. Analog-to-digital (A/D) converters are a prime example. In such cases, the simulation must be run on an analog simulator.

Performing digital simulations with SPICE requires special modeling because of the sheer size of digital devices. The obvious way to model a digital device is to describe it in a subcircuit using the actual transistor topology. However, this approach uses large amounts of computer memory and requires lengthy simulation times for even simple circuits.

A better way to simulate digital devices is using *threshold logic* theory. With this method, the input gate of the device looks at the two binary inputs as two analog voltages. The sum of each signal is multiplied by a weighing constant, and the resulting analog voltage is compared to a threshold value. If the sum is greater than the threshold, the output of the gate is treated as a logic one value; if the sum falls below the threshold, the output is treated as a zero. The weighing factors and threshold point are usually determined by polynomial algorithms. However, the algorithms are not easy to work with and are beyond the expertise of most SPICE users.

Obtaining SPICE Models

Fortunately, there is a large inventory of SPICE models available from many sources for several types of devices, including transistors, optoelectronics, and digital ICs.

All SPICE programs come bundled with a selected number of device libraries that range in size from a handful of popular components to several thousand parts. The number of parts provided is generally dependent on the price of the SPICE program, with the price increasing as the size of the libraries increase. Most SPICE programs also offer a number of optional component libraries which you can purchase at extra cost.

Many semiconductor manufacturers also have SPICE models for their catalog devices which they either offer for free or at a nominal charge. Outside the semiconductor industry, though, sources for SPICE models are more scarce. However, there are a few third-party vendors that provide SPICE models for specialty devices like microwave tubes, transformers, and acoustical filters — but expect to pay a premium for them.

The alternative is to roll your own. There are several publications and newsletters, mostly from SPICE software suppliers like MicroSim and Intusoft," that contain information and modeling tips for various types of electronic devices.

But no matter what your source of SPICE models, you must be aware of its possible limitations. Although functional modeling — modeling that uses voltage and current sources instead of individual components to simulate a device — is the most popular way to describe a device because the code is smaller and the simulation time is less, it may not be accurate for all applications. There is more than one way to describe a functional model, and some people are better at creating realistic algorithms than others. So do not take everything at face value. Fortunately, the situation is improving as better SPICE algorithms continue to be developed for both new and older devices.

Plotting SPICE's Output

A problem with SPICE, and most simulation programs, is that it generates a lot of data. In fact, too much data for most applications. All too often just sorting through the data can be a chore in itself. Fortunately, there are several ways to render the data into more useful form, the most efficient being graphics. Let's first look at the type of data a SPICE output file gives us.

Like the netlist, all SPICE output files are in ASCII, which means you can transport them from one computer to another without conversion or make a hardcopy of the file using any type of printer. Data is put into the output file in the same order that SPICE performs its operations. For example, the beginning of the file contains a netlist of the circuit under simulation, followed by the results of the dc analysis.

There are many ways to display SPICE output in graphics. The easiest is to use the .PLOT control statement that is built into SPICE. By adding a .PLOT statement to your netlist, you get a tabular listing similar to that generated by .PRINT plus a crude graphical representation of the values, as illustrated in Table 3-6. If you request more than one measurement within the same .PLOT statement, you get a multiple line graph which is handy for comparing two or more sets of measurements.

Table 3-6 Using the .PLOT Control Statement Built into SPICE You Can Make Crude Line Graphs of the Output[a]

```
******* 2/19/90 ******* IS SPICE  1.41 12/12/87*******20:30: 6*****

 BANDPASS FILTER

 ****     AC ANALYSIS                         TEMPERATURE =   27.000 DEG C

 **********************************************************************

  LEGEND:

 *: V(5)
 +: VDB(5)

     FREQ        V(5)

 (*)-------------   1.000D-01      3.162D-01      1.000D+00      3.162D+00  1.000D+01
                  - - - - - - - - - - - - - - - - - - - - - - - - - - - - - - - - - -

 (+)------------- -1.000D+01     -5.000D+00      0.000D+00      5.000D+00  1.000D+01
                  - - - - - - - - - - - - - - - - - - - - - - - - - - - - - - - - - -
  1.000D+03  4.773D-01 .          +   .    *       .             .             .
  1.100D+03  5.978D-01 .              .+      *    .             .             .
  1.200D+03  7.661D-01 .              .       +  * .             .             .
  1.300D+03  1.016D+00 .              .            X             .             .
  1.400D+03  1.406D+00 .              .            .   *    +    .             .
  1.500D+03  1.996D+00 .              .            .        *    .    +        .
  1.600D+03  2.481D+00 .              .            .           * .         +   .
  1.700D+03  2.182D+00 .              .            .         *   .      +      .
  1.800D+03  1.641D+00 .              .            .      *          + .       .
  1.900D+03  1.260D+00 .              .            .  *   +      .             .
  2.000D+03  1.013D+00 .              .            X             .             .
  2.100D+03  8.475D-01 .              .         + *.             .             .
  2.200D+03  7.296D-01 .              .     +   *  .             .             .
  2.300D+03  6.420D-01 .              .  +     *   .             .             .
  2.400D+03  5.743D-01 .              .+     *     .             .             .
  2.500D+03  5.206D-01 .           +  .     *      .             .             .
  2.600D+03  4.768D-01 .          +   .    *       .             .             .
  2.700D+03  4.404D-01 .        +     .   *        .             .             .
  2.800D+03  4.097D-01 .      +       .  *         .             .             .
  2.900D+03  3.834D-01 .     +        . *          .             .             .
  3.000D+03  3.605D-01 .  +           . *          .             .             .
                  - - - - - - - - - - - - - - - - - - - - - - - - - - - - - - - - - -
```

[a] If you request more than one measurement within the same .PLOT statement, you get a multiple line graph, like the one above.

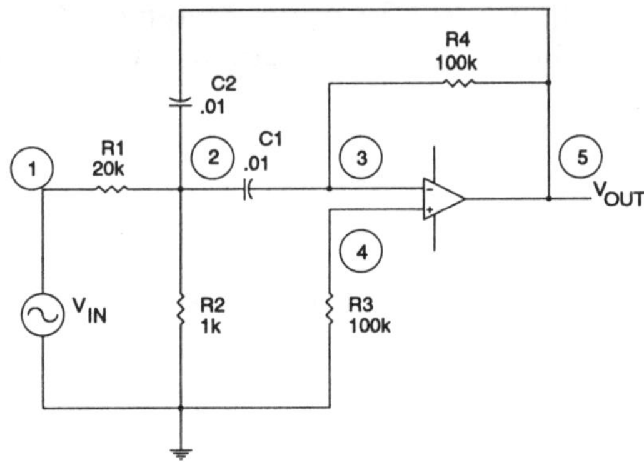

Figure 3-3 The PLOT command uses ASCII characters that print on any printer to produce this crude SPICE graphic.

However, .PLOT graphics leave a lot to be desired. While they are OK for cut-and-try laboratory design work, we do not advise taking them to your next board meeting. Fortunately, SPICE's tabular format is accepted by most database, spreadsheet, and graphics programs, including Lotus *1-2-3*. The best way to prepare a SPICE output file for data import is to use a line editor or word processor to remove everything but the desired tables. A Lotus graph describing Figure 3-3 is shown in Figure 3-4.

Choosing a SPICE Program

When choosing a SPICE program, there are several factors to consider, not the least of which is price. A SPICE package can cost anywhere from $100 to over $20,000.

The cheapest SPICE simulation package that we know of is *IsSpice* from Intusoft, which sells for $95. In comparison, the very popular *PSpice* program from MicroSim goes for $950. The reason for the large difference in price is due

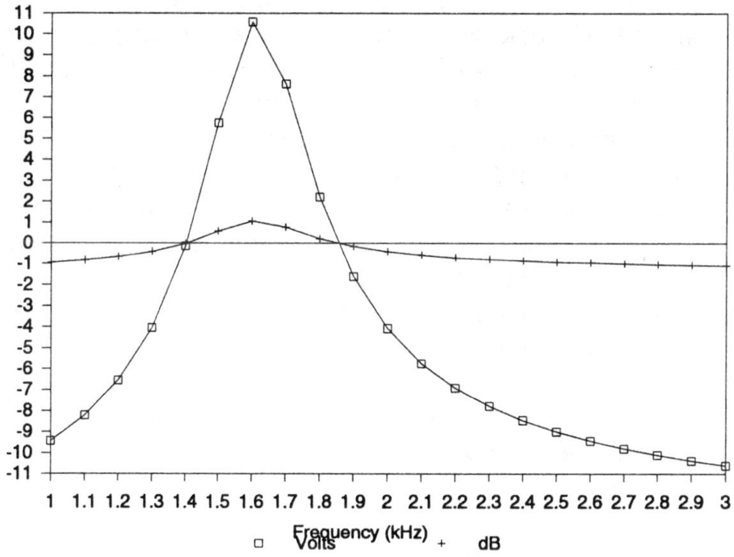

Figure 3-4 It is simple to import SPICE data into a Lotus 1-2-3 or <Excel> spreadsheet to get a decent plot.

mostly to the number of included device models. *IsSpice* comes with only a handful of transistor models, whereas *PSpice* comes with a transistor library that exceeds 1500 devices, plus numerous other device models.

Program prices also vary according to the type of PC you have. The prices of the *IsSpice* and *PSpice* programs mentioned above are for the PC-only version. If you want to use the power offered by a 286 AT or 386 PC, the program will cost more.

The type of PC also determines the size of the circuit the SPICE program can support. For example, the PC-only version of SPICE can only simulate circuits of 200 nodes or less (about 100 transistors, including subcircuits). For larger simulations, you need to buy more expensive versions of the program, which are machine specific.

Most SPICE packages offer a variety of utility programs that make SPICE easier to use. The most popular utilities are SPICE editors, Monte Carlo simulators, and output graphics processors. But get ready to raise the ante again, because these items cost extra for all but the most expensive SPICE packages. For example, a fully loaded *IsSpice* package goes for $790, and an equivalent *PSpice* package tips the scales at $1,750.

Yes, there are a lot of things to consider when buying a SPICE program. Fortunately, most SPICE makers have a demo package that lets you see how their program works before you buy, and we highly recommend that you examine the demo before putting down your money.

Chapter

4

Digital Circuit Simulation Using SUSIE

SUSIE, an acronym for "*S*tandard *U*niversal *S*imulator for *I*mproved *E*ngineering," is a popular software simulation program that graphically displays the timing and logic events of a digital circuit. Although SUSIE uses software to verify the design, the results are the same as if you had built the circuit and used an ultra-sophisticated logic analyzer to test and debug it.

Digital simulation programs differ considerably from analog simulation programs in that they disregard any reference to voltage or current. Everything in their world revolves around a pulse whose value is either black or white — there are no shades of gray.

Timing is critical to digital simulation. For example, let's take two gates that are cascaded together and compare them to a single parallel third gate. When a digital simulation program like SUSIE sees a situation like this, it immediately casts a suspicious eye because the propagation time through the serial gates is twice as long as it is through the single parallel gate. And at high clock rates this can create invalid logic states.

SUSIE comes in two versions that are different enough that you have to look at both. They are SUSIE 4.6 for the PC/AT and its compatibles, and SUSIE 6.0 for 80386 and 80486 systems. Our first look is at SUSIE 4.6.

Timing is crucial to digital simulation. For example, let's take two gates that are cascaded together and compare them to a single parallel third gate. When a digital simulation program like SUSIE sees a situation like this, it immediately case a suspicious eye because the propagation time through the serial gates is twice as long as it is through the single parallel gate. And a high clock rates this can create invalid logic states.

SUSIE comes in two versions that are different enough that you have to look at both. They are SUSIE 4.6 for the PC/AT and its compatibles, and SUSIE 6.0 for 80386 and 80486 systems. Our first look is a SUSIE 4.6

SUSIE 4.6

Like SPICE, SUSIE simulates the operation of the logic chips using algorithms. But rather than making voltages and currents a prime concern, SUSIE concentrates on hold time, pulse width, edge-to-edge transfer delay, and other propagation parameters.

SUSIE simulates the operation of the logic chips using mathematical models stored in a chip library. Chip models include algorithms for set-up and hold time, pulse width, edge-to-edge transfer delay, and other propagation parameters.

Chip libraries are divided into component types, with one library for TTL devices, one for CMOS, another for ECL, etc. SUSIE comes with the three above-mentioned libraries, plus two additional libraries for switches and other passive components. ALDEC's Model Builder Compiler (MOBIC) can be used to model chips that are not included in the libraries. You can also purchase extra libraries for devices like memory chips, microprocessors, gallium-arsenide (GaAs) logic, programmable logic devices (PLDs), and gate arrays, plus libraries for computer interfaces and selected industrial components like stepper motor controllers. Most of the optional libraries run between $800 and $2,000. But you can do a lot with the nearly 6000 devices that come standard with the basic SUSIE package.

Understanding SUSIE 4.6 by Example

To understand how SUSIE works, let's take an example. The circuit shown in Figure 4-1 is a divide-by-5 counter with a 50 percent duty cycle. Ordinary divide-by-5 counters produce an asymmetric output with a 20/80 timing ratio, but by using a negative edge-triggered flip-flop and half-

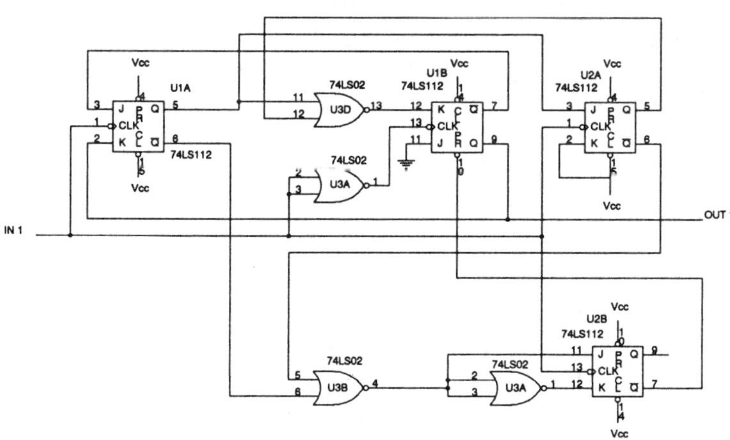

Figure 4-1 A divide-by-5 counter with a 50 percent duty cycle

state timing, we can make a divide-by-5 counter with a 50/50 timing ratio.

Three counter stages are needed to divide by five, with one driven on the inverted clock to compensate for the odd number of states. To achieve the sixth state needed to produce a 50 percent duty cycle, U1b must change states on the negative edge of the clock at $t = 2.5$, a change that can be forced by timely triggering of the flip-flop's preset function (pin 10).

Obviously this is not your typical digital design, and debugging a circuit like this in hardware can get quite thorny. Let's see how SUSIE handles the situation.

Writing a SUSIE 4.6 Netlist

SUSIE's design verification process is divided into three steps: design entry, test vectors, and simulation. Design entry is the step that loads and defines the section or sections of the circuit you want displayed for analysis on the simulation screen. SUSIE can show as much or as little of the design as you wish. You may zoom in on a single chip or display waveforms for every point in the circuit.

Design entry begins with a netlist (see Table 4-1). First the circuit's inputs, outputs, and components are identified by name and number. These labels are then entered into the netlist using the format:

```
sockets schematic label = device type
```

The arrangement of the I/O ports and parts is arbitrary, but these items must be the first order of business in the netlist. Next the connections between the devices are listed by component label and pin number, with a comma separating each entry. Signal inputs, outputs, and power supply connections are preceded by a slash. In this part of the

Table 4-1 Divide-by-Five Netlist Shown in Figure 4-1[a]

```
sockets u1=74LS112
sockets u2=74LS112
sockets u3=74LS02

/in1, u1/1, u2/13, u3/8, u3/9, u2/1      ;signal input
/out1, u1/2, u1/9                        ;divide-by-5 out
u1/3, u1/7
u1/5, u2/3, u3/11
u1/6, u3/6
u1/10, u2/7                              ;U1b preset
u1/12, u3/13
u1/13, u3/10
u2/5, u3/12
u2/6, u3/5
u2/11, u3/2, u3/3, u3/4
u2/12, u3/1
/GND, u1/8, u2/8, u3/7, u1/11
/VCC, u1/16, u2/16, u3/14, u1/4, u1/15, u1/14, u2/4, u2/15,
u2/2, u2/10, u2/14
```

[a] The order of the parts is arbitrary.

netlist comments can be appended to a line if the two are separated by a semicolon. Comments can be very helpful when making changes or corrections in a netlist — especially if the circuit is an existing design that is slated for an upgrade.

SUSIE's netlist is node oriented rather than component oriented. Each line in the netlist represents one node and all the components connected to that node. If the number of connections to a node is greater than one line can hold, the listing can extend to the next line by ending the first line with an ampersand (&) character. There is no limit to the number of lines you can use for a single node, provided the lines are in sequence and linked via ampersands.

A SUSIE netlist can be created using a text editor or word processor. Several schematic capture programs, including *OrCAD* and *Tango*, can also generate a SUSIE netlist directly from a screen schematic. However, not all schematic capture programs have SUSIE netlist algorithms installed. Fortunately, ALDEC (the makers of SUSIE) have an inventory of DOS conversion utilities that convert popular schematic capture hardware netlists into SUSIE netlist format.

Design Entry

The netlist is now ready for simulation. Unlike SPICE, no compiling is required. SUSIE netlists are ready to run from the get go, which makes editing of a SUSIE netlist less complicated than programs that require compiled netlists. SUSIE simply reads the ASCII version of the netlist, downloads the specified components from their libraries, and "builds" the circuit in computer RAM.

After SUSIE loads the netlist, you define what part of the design you want to look at by listing the pin number or name of the node you wish to display in the signal column (left side of the screen). For example, the screen in Figure 4-2 lists the input (IN1), output (OUT1), and selected pins of chips U1 and U2 for display.

Note that we selected no pins from U3 and only some from U1 and U2. The reason is that too much information can often be more confusing than not enough. The point is that, regardless of which nodes we choose to display, the entire circuit is tested during simulation and all questionable timing events (glitches, etc.) are brought to your attention whether they are slated for display or not.

Once a design entry has been created, it can be saved to disk and recalled for use later.

Test Vectors

For a design to become functional, a signal of some kind must be applied to one or more of its inputs using *test vectors.* A test vector is nothing more than a waveform that, when applied to an input, causes something to happen in the circuit. Test vector waveforms come in many sizes and shapes, and SUSIE can serve up an almost limitless variety of them.

Test vectors are created using either an ASCII text editor or SUSIE's built-in test vector editor. Test vectors may be stored in either a compressed binary format or one of three ASCII formats. The ASCII versions differ both in the format of the file and the way it is used.

The *line version* is formatted as strings of 1's (high) and 0's (low) that establish the timing and shape of the waveform. Each line in the file represents one waveform, and you may describe and load as many lines (waveforms) as you wish.

A *bus file* is similar to a line file, except that it is written in hexadecimal notation and is primarily used to define waveforms for an entire bus.

Waveform files, the third ASCII format, are written in ALDEC's proprietary high-level language, which can be used to create waveforms too complex for either of the above methods.

To use SUSIE's built-in editor, you simply place the cursor over the signal to be edited and manually key in the desired waveform. After the test vector is working satisfactorily, it may be saved to a line or bus file directly from the screen. However, test vectors created on the screen using ALDEC's programming language cannot be saved to file; ALDEC files can only be made using a text editor or word processor.

Once defined and named, a test vector file can be loaded into any netlist design entry. Although a test vector file may contain any number of waveforms, only one file may be loaded at a time.

SUSIE also has an internal 10-stage binary counter that can be used along with a test vector file as a square-wave signal generator or as a frequency divider for other test vectors.

Simulation

Circuit simulation is the third and final step in the SUSIE simulation process, and it is the step where you actually get to see what happens in the circuit as a result of the applied stimuli. It is also the day of reckoning, because here is where you learn whether or not your design will fly — and just how high.

The simulation mode provides numerous features that simplify design analysis. You can preset any design element to an initial logic state, search for timing glitches, and set breakpoints for incremental timing measurements.

Most simulations use the internal binary counter for the signal source, as we did for our test design in Figure 4-1. The counter has 10 stages (B0 through B9), each of which is half the frequency of the stage before it. The pulse width of B0 is adjustable between 10 picoseconds and 999 seconds. For our simulation, we set B0 at 15 ns and used B1 (30 ns) as the clock. The resulting display is shown in Figure 4-2, which we will now use to verify our design.

At power up (t = 0), the output of U1b (pin 9) is set high. On the first clock pulse, the input to pin 13 goes negative, causing the flip-flop to toggle and force pin 9 low and pin 7 high. This sets the stage for U1a to reset on the falling edge of the same clock pulse, which in turn causes pin 12 to go low and prevents further clock pulses from affecting U1b.

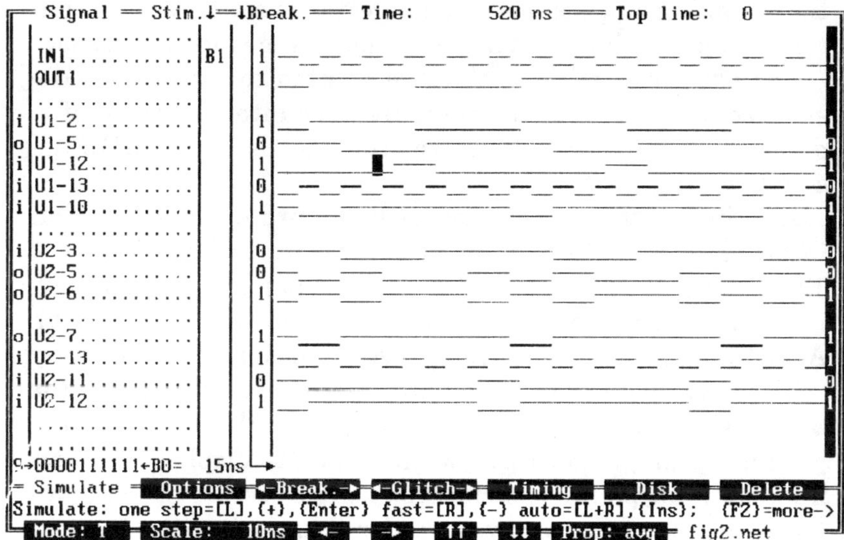

Figure 4-2 SUSIE simulation of the divide-by-five counter shown in Figure 4-1.

However, U2a continues to toggle, sending signals to U2b until its inputs reach a logical combination (pin 11 high, pin 12 low) that causes the inverted output to go low (pin 7) on the falling edge of the third clock pulse, forcing U1b's preset input (pin 10) low and output (pin 9) high. A half state later, U1b's new logic states force U2b's pin 11 input low, thus releasing the hold on U1b and enabling the counter to function as a normal state machine until this forced state is reached again. The resulting waveform at OUT1 shows us that indeed our design works as predicted.

Using SUSIE to Modify a Design

Unlike many analog and digital simulators, SUSIE does not need to compile the netlist before it can do a simulation, making it possible to change test vectors and circuit connections without having to make changes in the netlist.

For example, let's say that we wanted to modify our design so that it was a divide-by-3 counter instead of a divide-by-5 counter. This is easily done using SUSIE's on-screen editing features.

Looking at the timing screen in Figure 4-2, we see that in order to pull this off we have to force the break at t = 1.5 instead of at t = 2.5. We can test our theory by applying a custom-made test vector, which we will create in an ASCII file, to pin 10 of U1b to force this timely change.

The test vector can be created in several ways, but the best method for our situation is to use the ALDEC programming language. First we look at the display to determine at what points the changes have to occur, which gives us the shape of the test vector. For this application the test vector needs to have eight high pulses followed by four low pulses. The ALDEC equation for this waveform is (H8L4)100, where H8 specifies the eight highs and L4 specifies the four lows; the 100 indicates that the pattern in parentheses is repeated 100 times, giving us plenty of time to play with the circuit before running out of signal.

To perform the modified simulation, we load the original design into SUSIE and apply the test vector to pin 10 of U1b (Figure 4-3). Further analysis of the timing display also shows us that pin 12 of U1b needs to be tied high so that U1b is not affected by U2a. You could go into the netlist and change pin 12's connectivity, but SUSIE's test vector editor lets you assign IC pins or nodes to keyboard keys that manually override the design stimulus. In our case, we assigned the "a" key to pin 12, then toggled the input high for the simulation. The final results of this simulation, seen in Figure 4-3, show that our assumptions are correct and that the new design does indeed work.

The next step is to physically alter the circuit so that it generates the needed test vector itself. This can also be

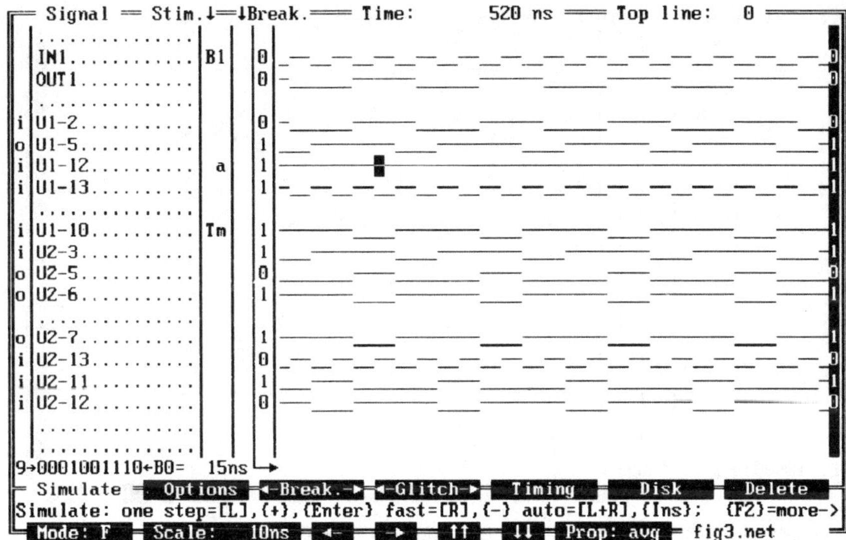

Figure 4-3 By manually changing the test vector on pin 10 of U1 and pulling pin 12 high, the circuit in Figure 4-6 becomes a divide-by-3 counter with a 50 percent duty cycle.

done in SUSIE using *connectivity markers* — screen notations that tell SUSIE to make a connection between like markers from one point to another. As it turns out, the only change we have to make in the original design is to move pin 5 of U3 from U2a to pin 9 of U1b using the "aa" connectivity markers shown in Figure 4-4. The rules for connectivity are: If you mark an input for screen connectivity, all previous connections to the input are severed; if you mark an output for screen connectivity, it serves the new source plus all the original sources.

We now disconnect the ALDEC test vector and run a new simulation. Once again, the simulation verifies the circuit changes. A revised schematic of the new divide-by-3 counter is shown in Figure 4-5, and its netlist is shown in Table 4-2.

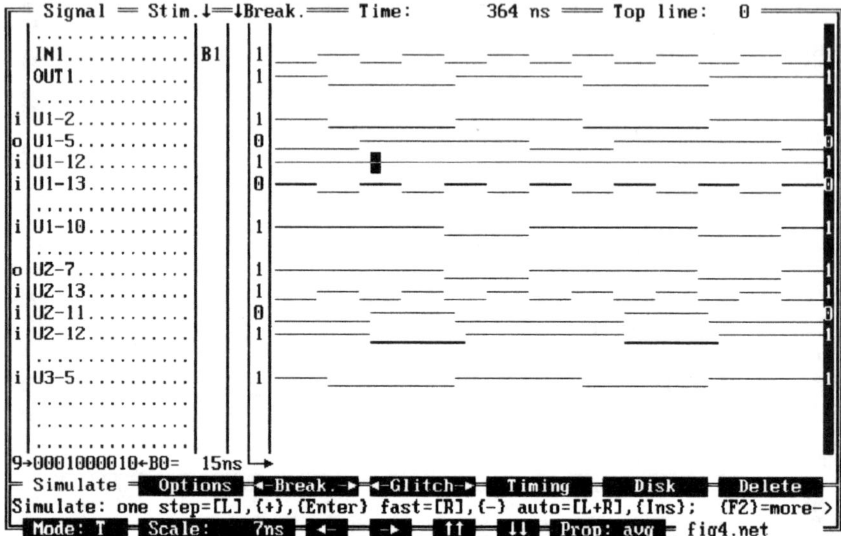

Figure 4-4 SUSIE simulation of the completed divide-by-3 counter.

Advanced SUSIE 4.6 Features

SUSIE 4.6 also has a host of advanced features that are used to pinpoint design defects, test for worst-case conditions, and input or output data to the real world. Here is a brief description of each.

Switches placed in the circuit can be opened and closed from the simulation screen, making "what-if" and alternate configuration designs a snap to verify.

The timing selection allows you to set chip propagation delays to any value for worst-case evaluation. You can also change chip technology (like go from LS to AS), save and load chip propagation data files, and include the effect of temperature and loading on the simulation.

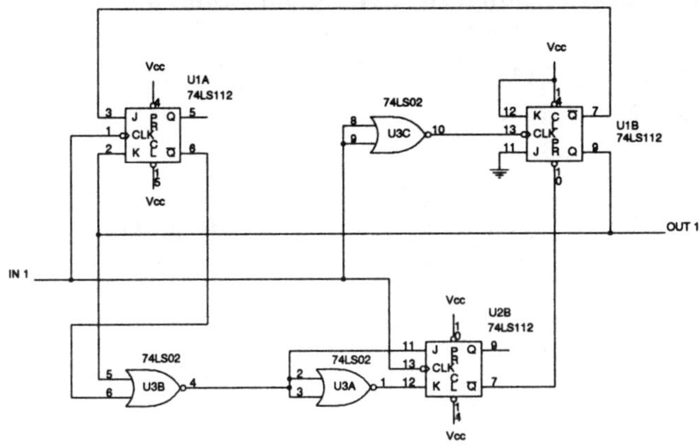

Figure 4-5 Schematic of divide-by-3 counter as modified in SUSIE.

Glitches that occur as a result of propagation delay, timing violations, or floating inputs are automatically displayed and pinpointed — even if the pin involved is not part of the simulation display.

Bus conflicts, where two or more outputs are vying for simultaneous use of the bus, automatically produce a warning message that can be used to pinpoint the areas in conflict.

Fault simulation that shows all stuck inputs and outputs is available as an option. To speed the lengthy process of fault simulation, SUSIE lets you divide the job among an unlimited number of unconnected PCs.

Hardcopy printout of screen simulations can be made using either a dot-matrix or laser printer. The logic analyzer option lets you feed test vectors from a hardware design (breadboard or PC board) to SUSIE for analysis and debugging.

Table 4-2 SUSIE Netlist of Divide-by-3 Counter

```
sockets u1=74LS112
sockets u2=74LS112
sockets u3=74LS02

/in1, u1/1, u2/13, u3/8, u3/9      ;signal input
/out1, u1/2, u1/9, u3/5            ;divide-by-3 out
u1/3, u1/7
u1/6, u3/6
u1/10, u2/7                        ;U1b preset
u1/13, u3/10
u2/11, u3/2, u3/3, u3/4
u2/12, u3/1
/GND, u1/8, u2/8, u3/7, u1/11
/VCC, u1/16, u2/16, u3/14, u1/4, u1/15, u1/14,
u1/12, u2/10, u2/14
```

SUSIE 6.0

As digital circuits grow ever more sophisticated, SUSIE 4.6 finds it increasingly difficult to keep pace. The newest and most advanced version, SUSIE 6.0, taps the power of Intel's 80386 and 80486 processors to provide the most comprehensive simulation results yet.

Like SUSIE 4.6, SUSIE 6.0 simulates the operation of the logic ICs using mathematical models stored in a library. IC models include all the functional properties of the device (valid logic input and output states) plus timing parameters for set-up and hold time, pulse width, and edge-to-edge transfer delay. To use the timing parameters, however, you must purchase the optional timing module, which is $3,995 at the time of this writing.

There are different libraries for different types of components, with one library for TTL devices, another for CMOS, and so forth. SUSIE 6.0 comes with two TTL libraries, plus one library each for mechanical switches and passive components. ALDEC's Model Builder Compiler (MOBIC) allows you to model ICs that are not included in the libraries. You can also purchase extra libraries for ECL, CMOS, memory, and programmable logic devices (PLDs). Cost of the optional libraries is $795 each for functional only and $995 each for functional with timing. Libraries for XILINX and ACTEL FPGA programmable gate arrays are also available for $995.

A SUSIE 6.0 Example

To understand how SUSIE 6.0 works, we will again use an example. The circuit shown in Figure 4-6 is a two-input data selector that lets you toggle the output between two input signals using a logic select line. The circuit can be used to change the clock rate in variable-speed designs.

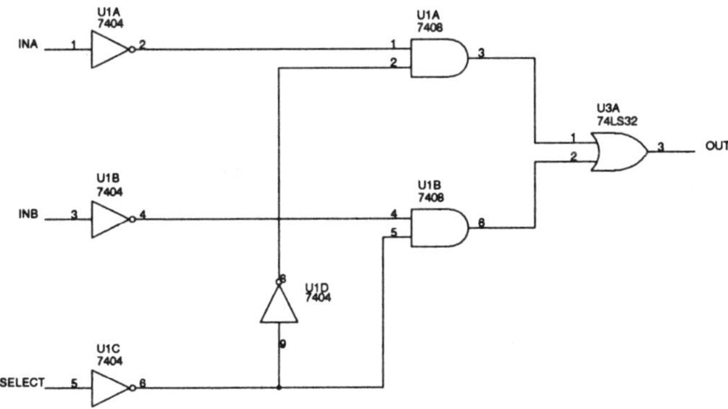

Figure 4-6 A two-input data selector.

The circuit is built around an AND-OR gate made up of U2 and U3. The output of U3 goes high when both inputs of either U2a or U2b are high; otherwise, the output is low. If we apply a low control signal to one AND gate and a high control signal to the other, only the gate with the high control line will pass the data input signal.

While the design is simple enough, it is not without problems. Propagation delay in the ICs distorts the rise times and introduces skewing that places a limit on the maximum speed the switch can handle.

However, clock speeds can be increased by changing to a faster technology, such as replacing the 74LS chips with 74F chips. To do that in a breadboard is not only a chore, it is a poor way of comparing one technology to the other. Let's see how SUSIE simplifies the job while increasing measurement accuracy.

SUSIE's design verification process is divided into three steps: design entry, test vectors, and simulation. During design entry, you load and define the sections of the circuit you want to display on the screen. SUSIE can display as much or as little of the design as you wish. You can zoom in on a single IC pin or display waveforms for every point in the circuit.

Writing a SUSIE 6.0 Netlist

Like all circuit simulation programs, SUSIE uses a netlist to define the circuit. But unlike many circuit simulators, which require special programming tools to create the netlist, SUSIE's netlist is in pure ASCII, which means all you need to make a SUSIE netlist is a text editor.

The SUSIE 6.0 netlist consists of seven building blocks. Heading the list is a file name that identifies the netlist. This is followed by an error-checking block that we will discuss later.

The actual netlist begins with a schematic of the design. (See Figure 4-6 and Table 4-3 for this discussion.) First, the circuit devices are listed by name in the Type field. Once a device is listed, it may be used more than once in the circuit, saving you the hassle of duplicating parts in the list. The order of the components must be alphabetical.

The matching of circuit components to device type is done in the Comp field. Whatever you place in the quotes is what the screen will show as the part value.

Next, the inputs, outputs, and other labeled signals are entered into the Sign field. The advantage is that you do not have to translate pin numbers into signal labels; this chart does it for you. Instead of trying to figure out what pin 1 of U1 is used for, SUSIE's simulation screen shows it as INA.

Next, the connections between the devices are listed in the Node field by component label and pin number, with a comma separating each entry. No special order is necessary, and comments (placed in quotes or after a semicolon) can be inserted in the netlist to help describe large designs.

A separate line is required for each node in the circuit (a node is any point where two or more pins or wires connect). However, if the number of connections to a node is greater than one line can hold, the listing can extend to the next line. There is no limit to the number of lines you can use for a single node.

After all the nodes are listed, you have to go back to the Qnty field and enter a check sum. The check sum consists of six values that are derived from elements in the netlist. The first value is the total number of netlist components, including device types and signal labels. This is followed by the number of nested devices. The number of signal labels and device types is then entered, followed by the number of defined delay lines. Finally, the check sum winds up with a node count.

The netlist shown in Table 4-3 is now complete and ready for simulation. A few schematic capture programs,

Table 4-3 Netlist of Two-Input Data Selector.

```
#File "Selector"

#Qnty  9, 0, 6, 3, 0, 12;

#Type
  1, "7404";
  2, "7408";
  3, "7432";

#Comp
  1, "U1",,1,C;
  2, "U2",,2,C;
  3, "U3",,3,C;

#Sign
  4, "INA",I,;
  5, "INB",I,;
  6, "OUT",O,;
  7, "GND",I,;
  8, "VCC",I,;
  9, "SELECT",I,;

#Node
  4.0, 1.1;

#Node
  5.0, 1.3;

#Node
  9.0, 1.5;

#Node
  1.2, 2.1;
```

Table 4-3 (continued) Netlist of two-input data selector.

```
#Node
  1.4, 2.4;

#Node
  1.6, 1.9, 2.5;

#Node
  1.8, 2.2;

#Node
  2.3, 3.1;

#Node
  2.6, 3.2;

#Node
  6.0, 3.3;

#Node
  7.0, 1.7, 2.7, 3.7;

#Node
  8.0, 1.14, 2.14, 3.14;
#EndN;
```

including *P-CAD* and *FutureNet*, can generate a SUSIE 6.0 netlist directly from a screen schematic.

Design Entry

Simulation begins by loading the netlist into the PC using a pull-down menu. The menus are tiled, that is, they overlay each other as you go deeper into the command hierar-

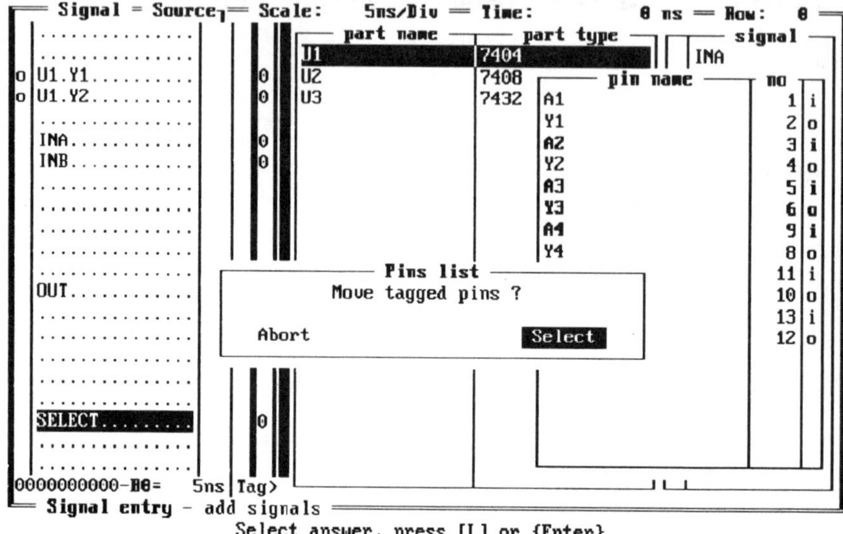

Figure 4-7 Pull-down menus are used to load and set up a SUSIE 6.0 simulation.

chy, and operate similarly to the pull-down menus of Microsoft *Windows*.

After the netlist is loaded, you define what part(s) of the design you want to look at by highlighting the pin numbers or node names from the Signal menus. After you have made your selections, the highlighted items are moved into the signal column on the left side of the display screen, as shown in Figure 4-7. The left-hand column identifies the signal as being input or output.

Note that we selected no pins from U2 and only some from U1 and U3. The reason is that too much information can be more confusing than not enough. However, the entire circuit is simulated when the program is run — no matter how many nodes are displayed — and all questionable timing events (glitches, etc.) are brought to your attention.

Of course, after creating a design and selecting the nodes to be displayed, you can save the set-up to disk and load it again later.

Test Vectors

For a circuit to function, you must apply a test vector. To do that, you first highlight the pins you want to simulate and then assign them test vectors. To simplify your work, SUSIE has an internal binary counter which can be assigned to input pins or input signals by entering them into the Source Field to the right of the Signal Field. The counter has 10 stages (B0 through B10), each of which is half the frequency of the preceding stage. The pulse width of B0 is adjustable between 10 picoseconds and 999 seconds.

You can also create test vectors using either an ASCII text editor or SUSIE's built-in test-vector editor. Test vectors may be stored in either a compressed binary format or one of three ASCII formats.

The line version is formatted as a string of 1's (high) and 0's (low) that establishes the timing and shape of the waveform. Each line in the file represents one waveform, and you may describe and load as many lines (waveforms) as you wish.

The bus file is similar to a line file, except that it is written in hexadecimal notation and is used primarily to define waveforms for an entire bus.

The waveform file consists of statements written in ALDEC's proprietary high-level language, and it can be used to create waveforms too complex for the other methods. In this format, SUSIE can serve up an almost limitless variety of test vectors.

To use SUSIE's built-in editor, you simply place the cursor over the signal to be edited and manually key in the desired waveform. After the test vector appears satisfactory, you can save it to a line or bus file directly from the screen using the Edit menu under the Main menu. However, test vectors created on the screen using ALDEC's programming language cannot be saved to a file; you can retain such files only by creating them first with a text editor.

After defining and naming a test vector, you can load it into any netlist design entry. Although a test-vector file may contain any number of test vectors, only one file may be loaded at a time.

Although the SUSIE 4.6 and SUSIE 6.0 test vector files appear to be identical, they are not. You must run a software conversion utility before SUSIE 4.6 test vector files can be used with SUSIE 6.0. SUSIE 6.0 files cannot be converted for use with SUSIE 4.6.

Simulation

Circuit simulation is the third and final step in the simulation process. The simulation mode provides numerous features that simplify analysis. You can preset any design element to an initial logic state, search for timing glitches, and set breakpoints for incremental timing measurements.

Most simulations use the internal binary counter for the signal source, as we did for our test design in Figure 4-6. For our simulation, we set B0 at 15 ns and used B1 (30 ns) and B3 (120 ns) as our input clocks.

To control the SELECT input, we must somehow toggle pin 5 of U1 between logic 0 and 1. This is done by assigning a keyboard character to pin 5; in our case, we selected the letter "a." Each time we press "a," pin 5 toggles between high and low. The keyboard control also takes precedence over any other signal that may be connected to that pin. For example, if we assign a keyboard character to INA, pin 1 assumes whatever state the keyboard forces it into. The rules for keyboard control are: If you mark an input for manual override, all previous connections to the input are severed; if you mark an output for manual override, it serves all inputs connected to that pin.

When the SELECT line is low, the control signal on U2b is high, which allows INB to appear at the output. When SELECT goes high, U2a's control line goes high and INA now appears at the output. The resulting display, which is created by toggling SELECT during the simulation, is shown in Figure 4-8.

Using SUSIE 6.0 to Modify a Design

Unlike many analog and digital simulators, SUSIE does not need to compile the netlist before it can do a simulation. That makes it possible to change test vectors and circuit conditions without having to change the netlist.

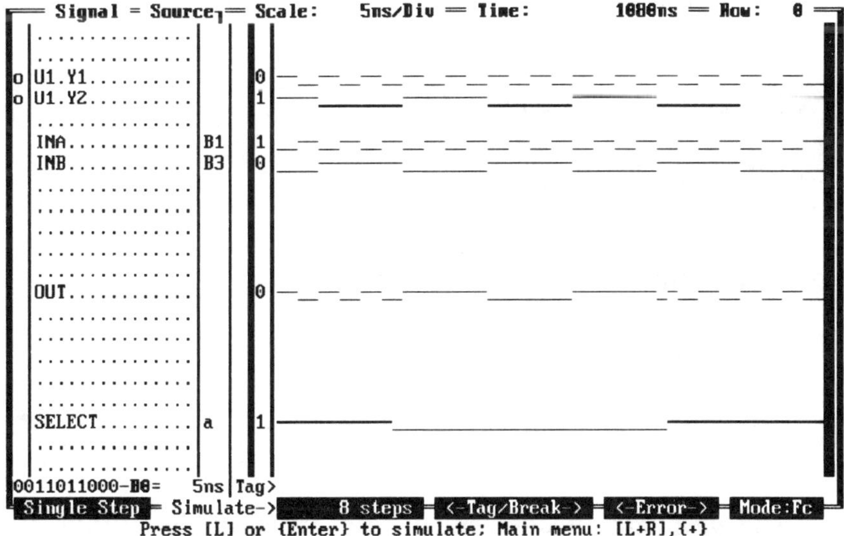

Figure 4-8 SUSIE 6.0 simulation of the two-input data selector shown in Figure 4-6.

For example, let's say that the propagation delay between INA and OUT is too long, and we wonder if going to another technology might improve the situation enough for the circuit to work. Let's say that we want to compare the 74LS chips to their 74F counterparts.

We do this by going into the Design Patching menu and selecting the Modify Chips Prop Specs command. At the prompt, we change the technology of our chips from LS to F, and simulate the design again.

The display shows that indeed the situation improves considerably. Instead of 35 ns propagation delay between INA and OUT, the delay time is reduced to a mere 13 ns.

SUSIE can also be used to change the memory mapping of ROMs and programmable logic devices (PLD). This makes it possible to design BIOS (Basic Input/Output Software) and ASIC (Application Specific Integrated Circuit) chips directly from the simulation program without having to go back into the logic table or netlist to make changes. All modifications made during the simulation can be saved to disk for later access.

Advanced SUSIE 6.0 Features

SUSIE 6.0 also has a host of advanced features that are used to pinpoint design defects, test for worst-case conditions, and input or output data to the real world. Here is a brief description of each (see Figure 4-9).

Design Patching allows you to set chip propagation delays to any value for worst-case evaluation. You can also evaluate the effect of temperature and loading on the simulation.

The propagation delay in lines connecting the output of one chip to the input of another can be modified to take into account changes in line lengths.

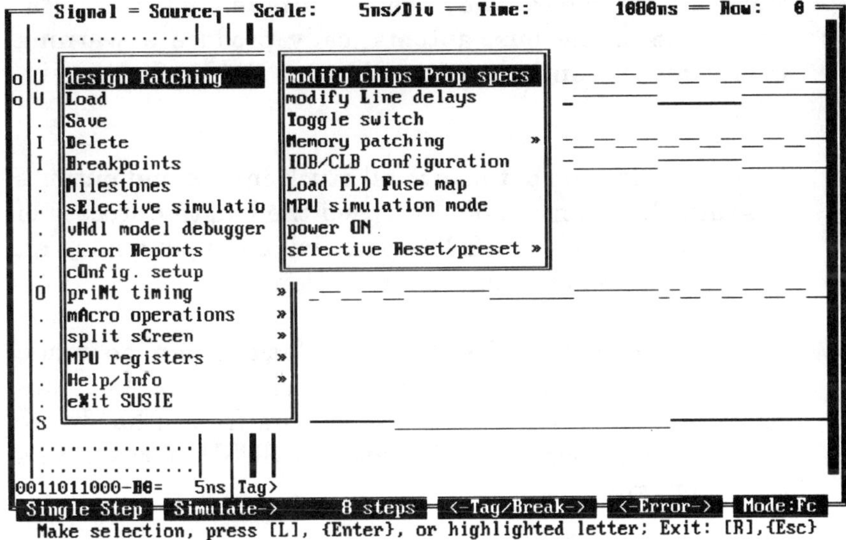

Figure 4-9 SUSIE 6.0 offers a host of advanced features.

Switches placed in the circuit can be opened and closed from the simulation screen, making "what-if" and alternate configuration designs a snap to verify.

MPU simulation allows for simulation of multiple microprocessors. Both the internal flag/register and hex file modifications are possible.

Breakpoints and milestones are used to identify and store design parameters for closer observation or later resimulation.

Glitches that occur as a result of propagation delay, timing violations, or floating inputs are automatically displayed and pinpointed — even if the pin involved is not part of the simulation display.

Bus conflicts, where two or more outputs vie for simultaneous use of the bus, automatically produce a warning message that can be used to pinpoint the conflicting signals.

Fault simulation that shows all stuck inputs and outputs is available as an option. To speed the lengthy process of fault simulation, SUSIE lets you divide the job among an unlimited number of unconnected PCs.

Hardcopy printout of screen simulations can be made using either a dot-matrix or laser printer. The logic analyzer option lets you feed test vectors from a hardware design (breadboard or PC board) to SUSIE for analysis and debugging.

Any sequence of SUSIE 6.0 operations can be saved in a macro file so that you may run the simulation again with a simple click of the mouse button.

Chapter

5

Printed Circuit Board Layout

After the circuit is verified, the next step is to lay out the printed circuit board. PCB software breaks down the making of a printed circuit board into four steps. The first is the physical placement of the components on the board itself. After the parts are in place, the copper traces that electrically connect the components together are laid down. Depending on the program, either or both of the processes may be automated.

The printed circuit board is then inspected for conformity with mechanical and electrical specifications using design rule-checking software. Violations are corrected using the editing tools supplied by the PCB program.

After the layout is verified, the final artwork is generated using a printer, pen plotter, or photoplotter. The art is used to make the actual printed circuit board. In addition, the PCB software produces production documentation that manufacturing needs for scheduling and inventory control.

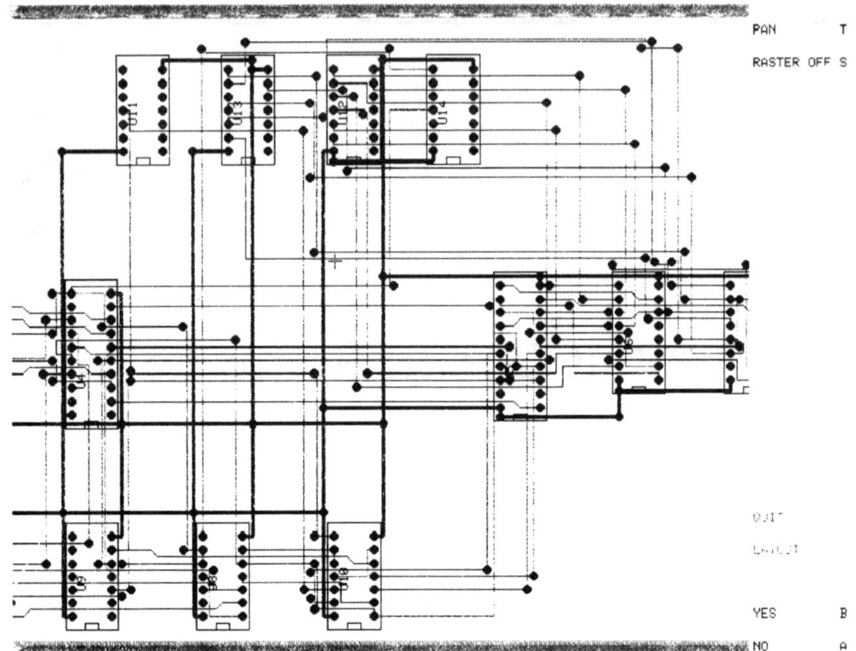

Figure 5-1 PCB programs can produce professional-quality PC board layouts.

Netlists

Although all PCB programs let you lay out a printed circuit board on the screen by hand, as you would using transfer tape and decals on a clear plastic base or copper-clad board, its real power is its ability to interpret data generated by a schematic capture program and convert it into components and traces.

The data is contained in disk files called netlists. A netlist is a grouping of related parameters that are extracted from the schematic. For example, a Bill of Materials (BOM) netlist lists all the components used in the circuit and their circuit references, whereas a wire netlist is a table of connections used to wire the various components together. Other netlists may include pin lists (listing only IC pin connections), design check reports, or circuit simulation files. There is no limit to the number of netlist types, and each schematic capture program has its own repertoire of netlists it can generate. These netlists are in ASCII format and therefore easily edited with a word processor.

Each schematic capture program also has its own way of organizing and formatting the netlists — rules that must be strictly adhered to if the PCB software is to interpret the netlist data correctly. The easiest transfer between schematic capture and PCB design occurs when the two are integrated in the same package. Next best is to buy the two programs from the same vendor.

However, there are a few schematic capture programs that have been around long enough that their netlist formats are recognized as *de facto* standards. Among them are *FutureNet, OrCAD,* and *Schema,* and it is not uncommon to find format translators for many popular schematic capture formats included with the PCB software.

Component Library

Central to the operation of the PCB program is a component library. The library is nothing more than a collection of device outlines like resistors, capacitors, ICs, etc. When a component is needed on the printed circuit board, the user or netlist file calls the specified outline from the component library and makes it available for placement on the board.

The PCB software has no idea of the function of the part it calls from the library. It is simply a mechanical device with size and shape. For example, the outline for a 74LS00 and a 74LS02 are the same, and the PCB software treats both parts identically. Most PCB programs support a wide range of device types, including optoelectronic and RF components.

Sometimes the component outlines are contained in specialized library modules, such as DIP devices or connectors. Other times the outlines are stuffed into one or two large library files. There is no advantage to one over the other, and the PCB programs are evenly divided on the issue. Additional components, particularly surface-mounted device (SMD) outlines, can often be purchased separately as either stand-alone libraries or as upgrades to the main library, depending on the program.

All PCB programs have a library editor that lets you create your own components and add them to the library. New parts can be created from scratch or by modifying an existing outline. Connectors are particularly problematic because of the many custom designs available. Unlike schematic capture library editors, most PCB library editors use primitives like circles and squares to build a new component, rather than have you draw the part using a fill-in-the-dot matrix.

Defining the Board

First the physical size and shape of the printed circuit board is defined. The size of the board determines how many components can be placed on the board, while its shape, to a large extent, determines the number of traces the board can support and their pattern.

Most printed circuit boards made today are plug-in cards that interface with a motherboard or backplane of some kind via an edge connector, as is the case with adapter boards made for the computer industry. These boards have a rigidly defined shape and size, plus defined connector placement and user control access areas. Popular printed circuit board outlines for computer use include the ISA (Industry Standard Architecture) and EISA (Extended Industry Standard Architecture) cards for the IBM PC/AT (and clones), VME, and Multibus. Other rigidly fixed printed circuit board outlines include Eurocards and DoD (Department of Defense) projects.

When working with a predetermined card size, the choices are narrowed from what is the optimum layout to what is the best that can be done with the space available. Of course, this constraint limits the number and types of chips you can work with, which is why ASIC (Application Specific Integrated Circuit) chips are gaining in popularity. ASIC chips can contain up to several thousand discrete components in a single package, thereby saving valuable board space.

Component Placement

Placing components on a printed circuit board can be tricky. Parts have to be positioned in such a way that the electrical connections can be made between each and every component. What seems like a functional layout can turn out to be a dead end — usually after you have invested a considerable amount of time trying to wire them together.

Fortunately, there are tools that can aid in the layout process, including automatic placement programs. However, automatic placement routines cannot achieve complete component placement. Problems with signal path length, clock skewing, and unconnectible traces are beyond the scope of today's automatic placement routines.

So whether you start from scratch or are modifying an existing automatic placement pattern, much of the layout work is done manually. Manual placement consists of dragging a part from one place to another using the mouse and/or the keyboard. Depending on the printed circuit

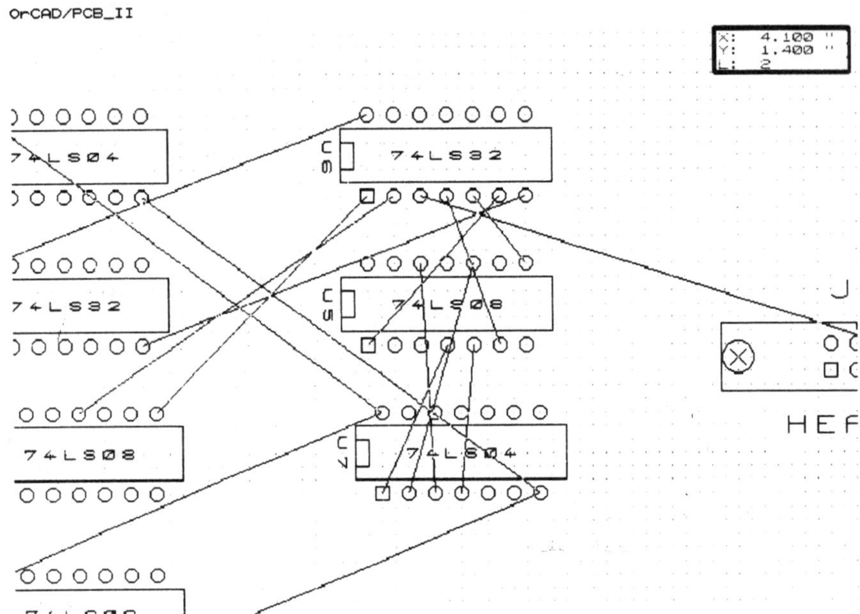

Figure 5-2 Ratsnests display the wiring connections among the components.

board software and whether or not the automatic placement program has been run, the parts may be located on the board itself, outside the board's outline, or stacked on top of each other in a single pile from which the components are removed layer by layer one at a time.

Manual Component Placement

Manual placement can be done either as a stand-alone job or interactively with an automatic placement program. With interactive placement, the computer gives suggestions for the best location of the parts, and the designer has the option of accepting or rejecting the placement on a part-by-part basis.

To assist you in placing the components, the PCB software provides an interconnectivity pattern on the screen called a ratsnest (Figure 5-2). A ratsnest is simply a jumble of skewed lines that show how each part is wired to every other part in the circuit. As a part is moved about, its attached ratsnest lines follow along with it, expanding and contracting in length like rubber bands.

The concept of the ratsnest is to associate the length of the ratsnest lines with the length of a completed trace. The object is to place the parts so that the ratsnest lines are as short as possible.

Unfortunately, as one component is moved to minimize the length of a group of ratsnest lines, another group of lines is lengthening to accommodate the part's new position. Further complicating the picture is that the ratsnest does not take into account overlapping lines which may make it difficult or impossible to route the connection given the component's present layout.

One way PCB programs enhance the use of the ratsnest is to display only the signal paths and hide the power supply lines. The reduction in the number of lines on the screen can make a seemingly impossible placement job manageable. Moreover, this tactic does not degrade the quality of placement because most multilayered boards (four layers or more) reserve the top and bottom layers for the power supply traces, making power supply routing independent of signal routing.

A few PCB programs, like *OrCAD* and Procad, use a force vector display that tells you in which direction a part should be moved for best connectivity (see Figure 5-3). But like the ratsnest, moving one part changes the force vector of another.

Zoom (scaling or magnifying image elements) is provided so that you can focus in on smaller areas of the printed

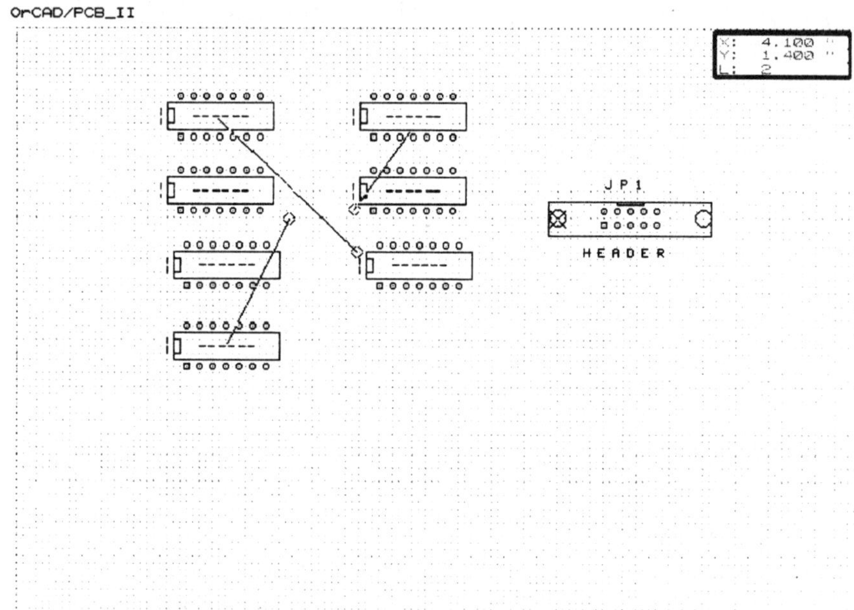

Figure 5-3 Force vectors are used to show in which direction the part should be moved to minimize trace length.

circuit board for accurate placement. Most programs have autopanning that automatically moves the board up, down, or sideways as you move the cursor beyond the borders of the present screen, but a few require manual scrolling of the screen.

Actual placement of the component depends on several factors, many of which are interactive and rely on the placement of a part before it. Let's look at these factors in their order of importance.

Foremost is the placement of connectors, switches, controls, and displays because they normally have rigidly defined locations. Connectors are the least flexible of all and are generally rigidly fixed as to size, type, location, and orientation. In most cases, the connector is an edge connector etched into the foil pattern. Exact placement of an edge connector is usually more critical than with a mechanical connector because of tighter alignment restrictions.

Some leeway may be possible with DIP switches and jumpers — provided they do not have to be user accessible from outside the card. User-accessible switches and controls are most often mounted along an outside edge of the board. The same is true for LEDs and other types of readouts.

After the immovable components are in place, the design should be viewed as a whole. Often the designer can see an association between certain components and the rigidly placed mechanical devices. This is particularly true of I/O (input/output) interface chips that link the board signals to the outside world via a parallel or serial cable connector. Whenever possible, these parts should be placed immediately next to the related connector or control.

Now comes the first round of abstract placement decision making. As before, a hard look at the overall picture may show the designer relationships between components, which can then be grouped together on the board. For example, it is a good practice to locate all the memory chips in one area of the board because they often use shared signal paths. Not only are the tracks shorter, but it is also easier for a service technician to locate components when they are grouped by function.

Automatic Placement

Automatic placement is a recent addition to printed circuit board layout programs. Automatic placement does not eliminate the problems of part placement, it simply makes them happen faster. When placement is done manually, a considerable amount of time is spent in deciding how the components should be grouped. A designer can easily spend a week manually laying out a printed circuit board with 200 ICs, whereas an automatic placement program could get the job done in less than a day. Instead of working on a design for a month to discover that the layout cannot be completely routed, the problem can be detected within a day or two, which cuts the time to market considerably. Even though the automatic placement layout may require more iterations to achieve complete routing, the total amount of time it takes to deliver a finished board is less.

However, automatic parts placement is in its infancy, and only a handful of programs include automatic placement — and those that do do not do it very well. The problems are easily understood when examining the way automatic placement programs work.

The automatic placement program is built on an algorithm that calculates the routability of the parts for a given layout pattern. By examining several different layout possibilities and comparing them, the program comes with a "good" placement layout that it considers to be maximally routable. However, a "good" placement to one algorithm may be unacceptable to another, and there are several different placement algorithms currently in use, each with its own merits.

A few of the simpler placement programs use overall connection length for its kernel algorithm, the simplest being the *straight-line distance algorithm*. With this algorithm, the merit of the placement is measured by adding up the lengths of its straight-line, point-to-point interconnections and choosing the layout that results in the smallest sum. Unfortunately, the connecting lengths are diagonals, which does not reflect the real distance the trace may have to travel.

An extension of the straight-line algorithm is the *manhattan distance algorithm*. Again the merit of placement is measured by adding up the lengths of the point-to-point interconnections and choosing the layout with the smallest sum. However, instead of using diagonals to connect the points, the manhattan distance algorithm uses perpendicular, horizontal, and vertical routes — as one would expect when navigating the streets of Manhattan to get from point A to point B. This algorithm more closely represents the actual traces on a printed circuit board, which are almost exclusively horizontal and vertical traces connected together at the corners with vias.

Sometimes restrictions are placed on the distance algorithm to fine-tune for different designs. For example, the placement program may be written to ignore very long traces. The reasoning is that any layout is bound to yield a few long traces which are inherently difficult to route no matter how hard the program tries, and the best layout can be more quickly found by concentrating on the shorter traces. Or very short traces may be ignored so that the placement program can concentrate its efforts on minimizing the number of long traces and maximizing their routability.

Unfortunately, routability does not correlate well with interconnect length. Although the algorithm has the advantage that it is easy to calculate and therefore runs very fast, it tends to bunch the components very closely together in the middle of the board. No provisions are made for specifying space between the parts, thereby limiting routability by not allowing enough space between parts for the traces and forcing the designer to manually separate them.

One solution is provided by algorithms based on density. This approach divides the board into boxes and calculates the predicted density of each box. The parts are moved about so that the density of each box does not exceed a routable limit. The weakness of this approach is the assumption that connections take up a certain amount of predetermined space, which is not always true. Small placement differences that allow an additional 16 signals to be connected without vias will have a significant impact on density and overall routability.

Another popular placement algorithm is known as *crossing count.* The theory here is that when traces cross, a via is required. Because vias take up more room and layers than unbroken traces, the program predicts that the layout with the lowest via count is the most routable and therefore the best. The problem is the program does not work well with complex layouts, generating a layout that gives long, snaking traces preference over shorter via routes.

While none of the placement algorithms promise total automatic placement, they are ideal for initial placement of the parts on the board. The layout may not be perfect, but it gives the designer a place to start — and it does it in a lot less time than he or she could do it manually.

Initial Placement

Anytime after the connectors and other immovable components are fixed in place, the automatic placement program can be run. The placement program is divided into two steps: initial parts placement and placement improvement. Ideally, both these steps would be automatic, but in many cases only one is fully automatic, with the designer required to perform the alternate step manually.

Initial placement takes the parts from the netlist and places them on the board according to the rules established by the placement algorithm. Most placement programs go about placing the parts using one of three methods. All three require a defined board and parts matrix before they can work.

The shape and size of the board is defined by drawing its outline on the screen. If there are areas on the board that cannot be used for parts, such as areas set aside for special mechanical assemblies like disk drives or expansion cards,

Figure 5-4 Strict attention to automatic insertion equipment spacing requirements must be observed when placing components on a PCB. Photo courtesy of Panasonic Factory Automation Company.

they too are defined at this time using fill areas that prevent parts placement within their perimeter. The resulting board area defines the maximum number of packages that can be mounted on the board, and you need to make sure it agrees with the package count in the design netlist — otherwise it is back to the drawing board.

Generally, the board outline only has to be defined on the first drawing layer, but some programs require you to copy the outline to all the layers used in the making of the board. Once a board has been defined, it can be saved to disk for future use.

Next a parts matrix is laid down. The matrix is needed to maintain conformity with mechanical and electrical manufacturing requirements. For example, if the board is to be assembled using automatic insertion equipment, the matrix needs to comply with the stepping actions of the insertion jaws.

The first method used to place parts on the printed circuit board piles the components in the center of the board in a big heap. The placement algorithm then applies repelling forces to the parts, causing them to pull away from each other. Conversely, the parts can be placed outside of the board and attractive forces applied to draw them onto the board. The amount of repelling or attracting force is determined by the placement algorithm.

To better understand what is taking place, imagine that all the components are interconnected by springs, and that the strength of the spring is determined by the placement algorithm's design rules. When the components are released, the forces among the various springs cause the parts to bounce about until a balance of force is reached. Parts with strong bonds come together, while those with weak forces tend away from each other. The balance is achieved when the placement algorithm is satisfied.

The third method has the user assign placement priorities to the parts matrix. Connectors, switches, and controls are assigned first, followed by grouped parts like memory chips. The placement program is then run, and the placements are locked into place. The designer then identifies the next most important component from the list of those remaining, and runs the placement program again, which selects the most appropriate location for that part. The process is repeated until all parts have been located.

Unfortunately, this method does not take into account the size of the components and will allocate one part per matrix point. Hence, large parts will likely overlap parts on neighboring matrix points, requiring manual intervention.

Some placement programs use one matrix size for ICs and another for smaller parts like resistors and capacitors. While it is a step in the right direction, it still cannot prevent large ICs from consuming their smaller neighbors.

The job is complete when all the parts have been placed on the board. The next phase is trace routing. However, before attempting routing, either manual or automatic, examine the layout closely for obvious problems, such as components that may mechanically interfere with card guides and mounting holes. Placement density is another easily recognized problem. If the parts are packed too tightly, the board may not have enough space between them for all of the traces to be routed, thus forcing the traces to take long meandering paths that consume more board space than would have been used if the parts were more widely separated.

Placement Improvement

A few PCB programs have a routine called placement improvement that is used to fine-tune the printed circuit board layout after the parts are initially placed. The two most popular improvement techniques are pair swapping and logic gate reassignment.

Pair swapping seeks to improve part placement through the interchange of neighboring components. If the exchange improves the placement value (i.e., reduces the trace lengths), the change is made. Otherwise, the components are returned to their original positions and a different pair is tried.

Another excellent way to improve PCB layout is by swapping logic gates. As a rule, the gate assignments are made in the schematic capture step of circuit design, with most programs assigning the gates to a package in the order in which they were placed on the schematic. While this ordering is quick and painless at the front end, it is not very efficient, and considerable improvement can be made in parts placement if the gates are grouped for optimum routability.

Some of the better swapping routines also permit pin swapping, which allows you to swap pins with the same function within the same gate. For example, all four input pins of a quad NOR gate are interchangeable. Generally, the program checks the pin exchange and denies swapping of pins with different functions.

However, part and gate swapping tends to get caught up in local politics that cannot see the forest for the trees. As the distance between parts increases, the likelihood of a distant component becoming involved in a local pair swap decreases dramatically, especially as the process nears completion. One way to eliminate swapping gridlock is to use three-way or four-way interchanges. Instead of two parts exchanging positions, three or four parts are moved as a block, thus minimizing local gridlock. But like extended pair swapping, the computer time increases quickly as the number of simultaneously swapped parts increases.

Placement Editing

Placement editing is an inevitable part of designing a printed circuit board. Even the best PCB designers have to go back to the drawing board at least once after they discover that there is no way all the connections can be made.

As a rule, PCB software lets you do anything you want when it comes to editing component placement. Parts can be moved, rotated, swapped, or flipped for placement on the opposite side of the printed circuit board (generally used with SMD devices). A PCB program that does not support all these functions severely handicaps the user.

The better PCB programs also support block editing. Block editing is a very powerful layout tool because it al-

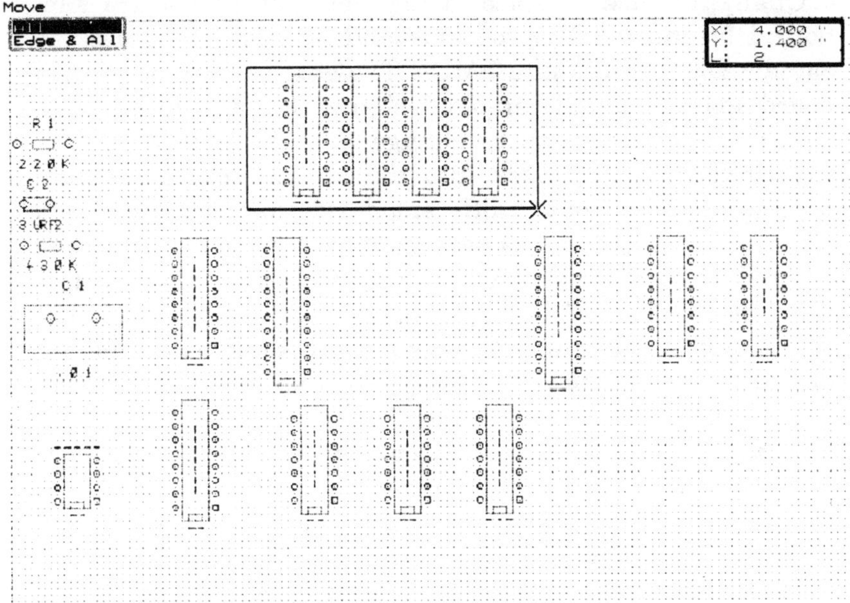

Figure 5-5 Block editing is a very powerful layout tool. The area inside the line is a block and is treated like a single object.

lows you to move several parts at a time without disrupting their relative positions (see Figure 5-5). Like single components, blocks may be moved, rotated, swapped, or flipped for placement on the opposite side of the printed circuit board.

A few PCB programs support macros. Macros are user-defined keystrokes and/or mouse clicks that are saved to memory so that they may be run again. Macros are most effective when shuffling large blocks of memory or interface chips around the board looking for an ideal location. Most macro software uses a macro recorder that remembers the sequence as you step through it, eliminating the need to write programming code.

Changing case outlines is another editing feature supported by most PCB programs and is most often used to change DIP devices into SMDs. The change may be to gain board space or update an old design with a new technology.

There are two ways to change a package outline. The first involves going back into the schematic capture program and calling out a different device name for the part. The second method has you change the package on the printed circuit board itself by calling out another outline from the component library. The problem with this is that the pinout of the new outline is not the same as the pinout of the original outline, and you are required to manually assign pin numbers and manually reroute all traces to the new package. This method is only good for layouts where the number of packages changed is either few or of the same type (memory chips, for example). There are exceptions, though. For example, both *PADS-PCB* and *Schema PCB* automatically convert pin assignments from DIP to SMD when the case outline is changed.

After the designer has made his or her placement corrections, the placement improvement routine can be run again to see if more subtle refinements can be made to the new layout. Here are other tricks the layout designer can call upon to improve layout and routability.

After the parts are permanently in place, it is often advantageous to reassign their identification nomenclature. It is a lot easier to find U2 next to U3 than it is to have it nestled between U38 and U39. Reassigning component identification gives order to the layout.

Trace Routing

The next phase is trace routing — the placement of copper tracks on the printed circuit board for electrical connections. Although you can route the traces manually, and will probably have to near the end of the job, most PCB programs have *autorouter* software that does the job for you.

Autorouter software uses the wire list netlist for its input. The autorouter then compares the placement of the parts on the board to the netlist connections and decides which trace should go where. It makes these decisions using a built-in algorithm.

Autorouters

The mainstay of autorouting technology is the *Lee* algorithm, created by C. Y. Lee in 1961. Sometimes the Lee router is referred to as a *maze router* or *flood router*. Another popular router with PCB software is the *Hightower* or *heuristic* router.

The Lee router is the more powerful of the two because it lets you enter cost functions that change its routing parameters. For example, if you want all the lines on the front of the board to run horizontally, you tell the router that the cost of a vertical line is very high, and the router will consider a vertical trace only as a last resort, trying all other options first. Other cost functions include maximum trace length, maximum number of vias (plate-through connections from one side of the board to the other), and trace density.

PCB programs usually refer to the Lee cost functions as strategy. A few examples of the Lee cost function jargon used in PCB software include normal, flexible, and extensive. Each term sets a limit on the trace routing, and you have to study the user's manual carefully to decipher their actions.

Lee routers have a very high trace completion rate, generally 90 percent or better. But get ready to pay the price for power. Lee routers are slower than most, and it is not uncommon for the router to spend hours or days on one printed circuit board.

The Hightower router uses a simpler algorithm with fixed parameters. As a result it runs much faster than a Lee router. What takes a Lee router an hour to do, the Hightower router can do in less than a minute. However, the completion rate is also less, particularly on complex designs.

A few PCB programs support both routers. Typically, you run the Hightower router first to get the bulk of the work done in a short time, then follow up with the Lee router to place the traces the Hightower missed.

Rip-Up Routers

The shortcoming of autorouters is they tend to box themselves in. Without the foresight to see that placing one trace will block the path of another, the program grinds to a halt with traces left unconnected.

The solution is to use a clean-up router that can get the offending traces out of the way. The two approaches are rip-up and shove-aside.

Rip-up, sometimes called rip-up-and-retry, routers are cheaper and easier to use and are sometimes included in the PCB program. Generally, though, they are sold as separate programs, ranging in price from several hundred to several thousand dollars. As with schematic capture programs, the rip-up program must support your PCB netlist format before it will work.

The rip-up router works by searching for unconnected traces and looking for a single trace that can be removed to make its completion. The offending trace is first ripped up, the blocked trace is routed, and the program proceeds to find a new path for the deposed trace.

The strength of rip-up routing is that it can achieve very high completion rates, with a good rip-router achieving 100 percent completion nearly every time. However, few rip-up routers can do the job unassisted. Most require the user to identify which trace is to be removed, after which the rip-up program attempts to make the new connections. Those that can do it alone are quite costly and very slow.

Shove-aside routers move traces rather than destroying them. They are most useful when a trace has a path to its destination, but does not have enough room for the track. However, only a couple of popular PCB programs offer a shove-aside router as an option, and the most popular shove-aside router on the market, *MaxRoute,* sells for a hefty $6000.

Trace Editing

After all automatic trace routing avenues are exhausted, the designer must resort to manual routing and editing. Manual editing is also used to fix design rule violations and to make design changes in printed circuit boards already in production. Several trace routing editing tools are available.

Route and delete are the most often used. Route takes many forms depending on the PCB program, but usually it is a straight line whose path is defined by a mouse. Although the trace is usually made up of straight segments with angular turns, a few PCB programs let you place curved tracks, like the kind needed for laying out analog circuits.

Delete is most often used during the final stages of routing to remove a trace blocking the way of another trace. Like traces that have yet to be placed, deleted traces are replaced with their ratsnest counterpart, lest you forget which traces remain unconnected.

Some PCB programs can modify traces without deleting them. This editor lets you push and shove the trace, place and delete vias, and move segments from one layer to an-

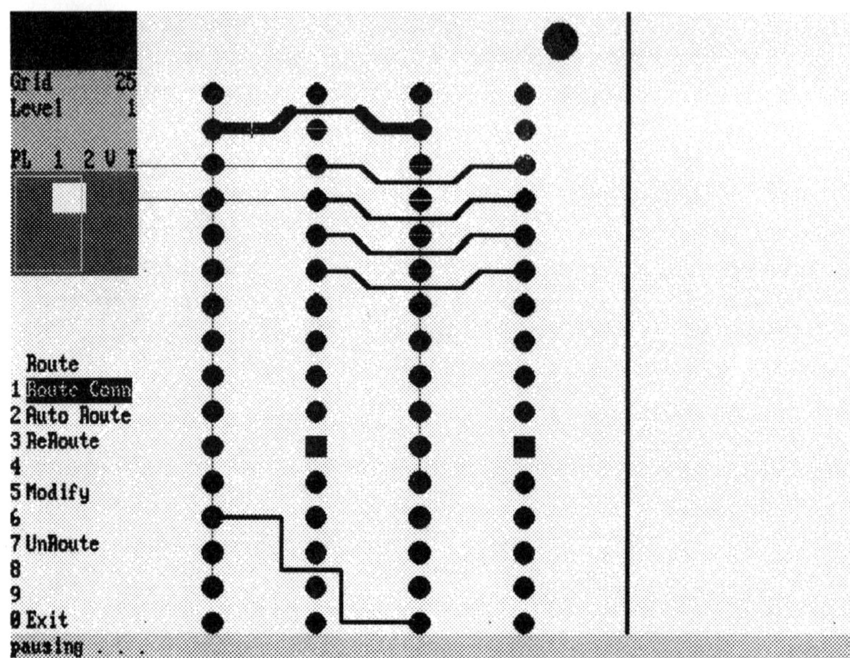

Figure 5-6 Necking down reduces track width for routing between solder pads.

other while maintaining the validity of the original connection.

The width of single traces can also be adjusted wider or narrower as the situation demands. A useful width function is necking down, which lets you narrow a short portion of a trace so that it may squeak between the pads of an IC without shorting (Figure 5-6).

Manual editing is also required for defining ground planes, protected zones, and most nomenclature. Find, search, and jump locate particular modules or reference designators on the printed circuit board.

Design Rule Checking

Design verification is a very important step because mistakes that pass through here end up as mistakes on the printed circuit board, which are costly to correct. The verification process checks to see that the tracks, vias, and pads have been placed according to a set of rules which you have established. The process is known as design rule checking.

The design rule checking software first loads the wire list netlist, then checks to see if all the nodes in each net are connected. If they are not, an error message is generated for each missing connection and placed in a design rule netlist file. The software also checks to see if there are any extra pins.

A check is then made of the traces, pads, and vias. Traces and pads that are too close together or touching are identified and marked for display. Most PCB programs highlight the affected area on the screen plus display an error message indicating the coordinates of the problem areas.

PCB Artwork and Output Files

The ultimate goal of a PCB program is to produce artwork than can be used to make a functional printed circuit board. The output is placed in netlist files of three different types: artwork, drawings, and documentation.

First and foremost are the artwork files, which contain the trace patterns for the printed circuit board. Each layer of the board has its own separate trace pattern; multilayer patterns are aligned using placement holes drawn on each layer at the beginning of the PCB design process.

Artwork can be printed out using a dot-matrix or laser printer, pen plotter, or photoplotter. Photoplotters, which

Hardware Requirements

Like schematic capture programs, PCB software requires high-resolution screen graphics. Although a few PCB support IBM's old CGA and EGA resolutions of 640 × 350 or less, no less than VGA graphics with 640 × 480 resolution in at least 16 colors should be considered for PCB layout. The reasons are twofold.

First and most important is that when working with a software package that can locate and place an IC or a trace with a tolerance of 1 mil (one thousandth of an inch), trying to display that on a screen of less than VGA resolution is an exercise in futility. And with high-density designs, even superVGA screen resolutions of 800 × 600 and 1024 × 768 are pushed to the limit. Most VGA boards support at least one superVGA mode but require a multisync monitor to display resolutions beyond VGA.

Second, the PCB software uses color to identify the numerous drawing layers needed in the layout of a printed circuit board. Beyond the obvious copper layers (which can number up to 64), there are additional drawing layers used for nomenclature, solder masks, and vias.

PCB design software is also more memory intensive than schematic capture programs, with many programs requiring or supporting LIM 4.0 expanded memory. When working with boards containing 50 or more chips, expect to have at least 1.5 MB of RAM installed.

Hard disk requirements are the same, with most PCB programs requiring 2 MB to 5 MB of hard disk space. A math coprocessor is also recommended, as is a mouse.

are akin to phototypesetters, are the preferred choice because of their high resolution and accuracy. Unfortunately, photoplotters are very expensive to own and operate, and most companies send their artwork files to a photoplotting service that converts them into photographic transparencies suitable for PCB manufacturing. Gerber format photoplotting produces the most accurate PCB transparencies and is supported by virtually every PCB program and photoplotting service.

Other artwork produced by the PCB software includes solder masks (that apply an epoxy film on the printed circuit board prior to flow soldering to prevent solder bridges from forming between adjacent traces) and silkscreen mats for board nomenclature. Many PCB programs also support Excellon's N/C drill file format that robotic drilling machines use to locate and drill properly sized holes in the printed circuit board.

The PCB software also produces a flurry of documentation needed for manufacturing. They include schematics, parts placement drawings, and engineering change orders (ECO), among others. Most of the documents can be printed on a dot-matrix or laser printer.

Last but not least is the back annotation netlist that contains all the changes made to the design during the PCB layout procedure. Included are logic gate reassignments, pin swaps, and revised component identification, all of which must be reported back to the schematic capture program for correction.

Chapter

6

Evaluating Circuit Design Software

As discussed earlier, circuit design programs are software packages that let you go from a design concept to schematic capture to printed circuit layout. However, there are several schematic capture and PCB layout programs available, and they differ considerably in both price and performance.

In this part of the book we look at 28 popular circuit design programs and evaluate them for price, performance, and features. Included are schematic capture, PCB layout, autorouters, and interface utilities. Only fully functional versions of the software were considered for review. Demo or evaluation copies of the software were not permitted or used. However, all the vendors reviewed here have a demo disk that you can obtain (usually for free) if you want to gets a hands-on feel for the program before you buy. The software reviewed is listed in Table 6-1.

Prices of the programs range from $99 to $2,395. The prices quoted, which are list, have been firm for the last

Table 6-1 Circuit Design Software Products Reviewed

Program Name	Program Type	Company
CapFast	Schematic Capture	Phase Three Logic, Inc.
DC/CAD	Integrated System	Design Computation, Inc.
EE Designer III	Integrated System	Visionics Corp.
FutureNet-5	Schematic Capture	Data I/O Corp.
HiWIRE II	Integrated System	Wintek Corp.
Intelligent Menu System (IMS)	Interface	Voltec, Inc.
ISIS Supersketch Plus	Schematic Capture	Labcenter Electronics
OrCAD/PCB II	PCB Layout	OrCAD Systems
OrCAD/STD III	Schematic Capture	OrCAD Systems
PADS-PCB	PCB Layout	CAD Software, Inc.
PCB II	PCB Layout	Labcenter Electronics
PCBoards	PCB Layout	PCBoards
PCRoute	Autorouter Design	RGH Software
ProCAD	Integrated System	Interactive CAD Systems
Protel-Autotrax	PCB Layout	Protel Technology, Inc.
Protel-Easytrax	PCB Layout	Protel Technology, Inc.
Protel-Schematic	Schematic Capture	Protel Technology, Inc.
Schema III	Schematic Capture	Omation
Schema-PCB Layout	PCB Layout	Omation
Schema-Quik	Schematic Capture	Omation
SuperCAD	Schematic Capture	Mental Automation, Inc.
Tango-PCB Plus	PCB Layout	ACCEL Technologies, Inc.
Tango-Route Plus	Autorouter	ACCEL Technologies, Inc.
Tango-Schematic	Schematic Capture	ACCEL Technologies, Inc.

several years and are not expected to change any time soon. However, there is normally a nominal charge for upgrading the software as new versions are released.

Because many of the high-end products bring a premium price, we opted to look at their budget-priced versions when available, which the vendors refer to as their university or student packages. They do everything the professional version does, but on a smaller scale. Savings of up to $500 can be realized by going this route. When there was a difference between the professional and student version of the software, we looked at both.

Fully Integrated System Software

Four of the programs are fully integrated packages that come with both schematic capture, PCB layout, and an autorouter. *EE Designer III* further provides analog and digital circuit simulation to give designers an idea of how the circuit performs before actually building it, which means that any one of these packages is all you need to have a complete circuit design system. However, this does not mean that you are totally locked into what the package provides. Software options like a rip-up router or extra netlist formatting utilities are generally available. Prices range from $495 to $1,995.

Because schematic capture and PCB layout software must be evaluated differently, the review of these four programs is split in half. In the following paragraphs, we tell you how the two operations of schematic capture and PCB layout stack up against their stand-alone counterparts. The four integrated systems programs we will be referring to are *DC/CAD, EE Designer III, HiWIRE II,* and *ProCAD.*

Schematic Capture Software

Nine of the programs are schematic capture software that let you draw schematic diagrams using your personal computer, thus reducing the time it takes to produce a schematic and improving its accuracy. In each case, we have selected software that is tied to a related PCB layout program so that you are not caught buying a package that cannot talk to a PCB layout program to finish the project. Prices range from $99 to $895, with five of the nine listing for $495.

The number of parts contained in the component library, the single most important element of a schematic capture program, range from 350 to over 11,000 devices. All but one component library is made up of several library modules, which allows you to delete unused modules to save hard disk space. Library modules are grouped either by generic type (such as TTL or CMOS) or by manufacturer (Intel, Zilog, etc.). Only *EE Designer III* puts all its parts in a single library module.

Seven of the nine stand-alone programs and three of the integrated programs mentioned above (the exception is *EE Designer III*) have autopan, which automatically scrolls the worksheet when the cursor touches the edge of the screen. However, only a few are smooth scrolling; the rest have a jump pan instead, and *ProCAD* will not autopan from the top down. The remaining three force you to pan manually using a dedicated pan command, scroll bars, or by manipulating the zoom controls.

Seven programs, two of which are integrated systems, have an on-line help screen. Macros are supported by TK. All support both flat and hierarchical drawings.

Two programs, *FutureNet-5* and *HiWIRE II*, will not let you disable the ortho drawing mode, making it impossible to draw diagonal lines or wires. Six of the programs, including three integrated system packages, let you draw 45-degree diagonals while in the ortho mode.

Four of the programs, two of which are integrated systems, have a hardware security device called a *dongle* that must be plugged into the parallel printer port before the program will load. The printer then plugs into the dongle and operates as though the dongle was not there. *ProCAD* uses a special software protection scheme that lets you load the program on a hard disk just one time.

All except *ISIS Supersketch* generate netlists that can be used to interface the schematic capture file with PCB layout software. However, the netlist formats vary widely, depending on the program's needs. The integrated PCB layout packages generally have little need to support more than one format because everything is done in-house. Schematic capture programs that are tightly linked to a PCB software product also tend to limit their netlist files to a narrow range of formats. Five programs output 14 or more formats, and you will find scattered support for *AutoCAD* DXF and PostScript files.

All netlist files are in ASCII format and are easily edited using a word processor. Of importance to most circuit designers is a Bill of Materials netlist, or BOM. The BOM lists all the components used in the circuit and can be used to create purchase orders and inventory control.

Hardcopy printout also varies considerably for the nine programs. All have at least minimal support for a dot-matrix printer that gives you a fair to excellent rendition of the drawing. Three let you fit the drawing to the page size, and all but *SuperCAD* support a plotter. The amount of printer or plotter control you have from the software also varies considerably, ranging from no control to full control.

PCB Layout Software

There are 12 programs that can do PCB layout, four of which are the integrated systems mentioned above. All but two can read and extract data from a schematic capture program. They are the Labcenter Electronics *PCB II* and Protel's *Protel-Easytrax*. The two exceptions were included because both can be upgraded to a better product that will accept their PCB layout output netlists and because both showed excellent features at a good price.

Prices of the programs range from $99 to $1,495. However, the less expensive programs usually require optional programs like an autorouter to make them fully functional. In the case of *PADS-PCB* and *Schema PCB Layout,* the total comes to $2,395. However, the average price of a fully configured system is between $1,300 and $1,600. Buying the schematic capture software that it takes to make six of the eight stand-alone PCB layout programs into complete circuit design systems adds another $500 or so to the cost. But before you throw up your hands, let us tell you that you can put together a complete circuit design system, with schematic capture and an autorouter, that is suitable for hobby projects for under $300.

The link between the schematic capture and PCB software is critical in many ways. For example, three of the programs cannot extract device outlines from the schematic netlist, forcing you to get the parts from the PCB library by hand. Forward and back annotation is also important, because it lets you keep up with changes made to both the schematic or PCB layout. This is particularly important when swapping gates or reassigning device references. Four programs have no annotation capabilities.

Of those programs that can extract device outlines from the netlist, all but two demand that you define the board outline before the parts can be placed. The two exceptions, *OrCAD/PCB II* and *TK,* have you define a working area that may or may not have any resemblance to the finished board. *PADS-PCB* and *Schema PCB Layout* require both a board outline and a parts matrix before they can import parts from a netlist. Board outlines can be saved in a file, making it a simple matter of loading a stored outline from the database when it is needed. Most programs support boards up to 32 × 32 inches; one can only support 9 × 13 and two can support boards of up to 64 × 64 inches.

Five of the programs, including *EE Designer III* and *DC/CAD*, have an automatic parts placement routine that not only extracts the devices from the netlist, but places them on the circuit board. As a rule, the placement is not as good as if you did it yourself, but it sure saves a lot of time. All but one, *PCBoards*, let you flip parts to the opposite of the board for placement of SMDs (surface mounted devices). All the programs with the exception of *PCB II* and *Protel-Easytrax* have a ratsnest to aid you in your placement and track routing, but the ratsnest may not always be available for all phases of parts placement. Two programs, *OrCAD/PCB II* and *Tango-PCB Plus*, have both a ratsnest and force vectors.

Six programs, including three integrated system packages, come with an autorouter that automatically draws your tracks for you, and four offer an autorouter as an option. Only *PCB II* and *Protel-Easytrax* force you to do it by hand, because neither can read a netlist. However, *Protel-Easytrax* has a pad-to-pad autorouter that makes the job easier. The number of copper layers support ranges from two to 64, with several of the mid-priced packages supporting six to eight layers. All but one lets you select the color of the layer. All let you change pad and via size, shape, and orientation, and about half let you use blind and buried vias. Nine of the twelve, including all the integrated system packages, have a design rule check function that catches broken tracks and spacing violations.

A hardware security device, the dongle, must be plugged into the parallel printer port before four of the stand-alone programs will load and run on the PC. Two of the integrated system packages use a dongle, and *ProCAD* is copy protected, which allows the program to be loaded onto only one hard disk at any time.

Hardcopy printout varies considerably among the programs, but all support a dot-matrix or laser printer and Hewlett-Packard or Houston Instruments plotter, but you may have to pay extra for it. Gerber and Excellon N/C drill files used for automated manufacturing of circuit boards is supported by all but one program, but again it may be at extra cost.

A tabulation of these and many more features can be found in the Appendix.

Chapter

7

DC/CAD, Design Computation

DC/CAD is a complete schematic capture and PCB layout circuit design program that sells for an astonishingly low $495. Even more amazing is that this low price includes automatic parts placement and a rip-up autorouter. It is easy to use and very versatile. However, it is not the easiest program to learn, nor is it the fastest.

The program requires 640K of system RAM and can use up to 3 MB of LIM memory if available. A hard disk is required. Happily, a hardware protection key is not needed to run the program and the program is not copy protected.

Learning to use *DC/CAD* is not easy. It has a rather unusual menu-driven command structure that is unlike anything else in the industry. But once you get the hang of it, using the program is really quite simple.

Although *DC/CAD* uses the same screen to do both schematic capture and PCB layout, the transfer of data between the two is not seamless. Many of the circuit design functions — including netlist generation and autorouting — are done in DOS, which means you have to quit the program frequently during the design process.

DC/CAD has an autopan that moves the screen by three-fourths of the viewing area when the cursor bumps against an edge, and the autopan may be disabled. Zoom is infinitely variable and can be used on the fly while placing parts, nomenclature, and wires or tracks.

There is a macro recorder for both schematic capture and PCB layout that remembers the commands you make and replays them on demand. Macros save a great deal of time when drawing or placing memory arrays and other repetitious patterns.

A help command that provides detailed information on all aspects of the program is provided, but it is arranged as an index rather than being context sensitive. If you don't need help but just a gentle reminder of the command's function, there is a command description line at the bottom of the screen that briefly describes the function of the command selected.

Schematic Capture

When starting *DC/CAD,* the program defaults to the schematic capture mode.

The component library consists of about 2000 parts distributed among 11 libraries, all of which are active at the same time. However, many of the devices in the libraries are identical, differing only in name. For example, the contents of the 74LS library is identical to that of the 74AS library — which is identical to the 74ALS library. A count of *different* device types reveals a true library size of only 150 devices, a remarkably small number.

Fortunately, making new components is as simple as modifying an existing part using the drawing elements found in the schematic capture mode. Parts can also be made from scratch using this method. No special library editor is required, but everything on the screen becomes part of the new component when saved. While this is an excellent way to create subcircuits, it requires clearing the screen of everything not related to the new part. Design Computation also has a bulletin board which has many schematic and PCB devices made by *DC/CAD* users that are free for the downloading.

Parts are called up from the library by typing their name at the ADD SYMBOL prompt. Each part must be typed in separately because *DC/CAD* does not support multiple placements, and you will need to use the reference manual for the part numbers. Components can be rotated or mirrored only after they are initially placed; the degree of rotation is infinitely variable between 0 and 360 degrees.

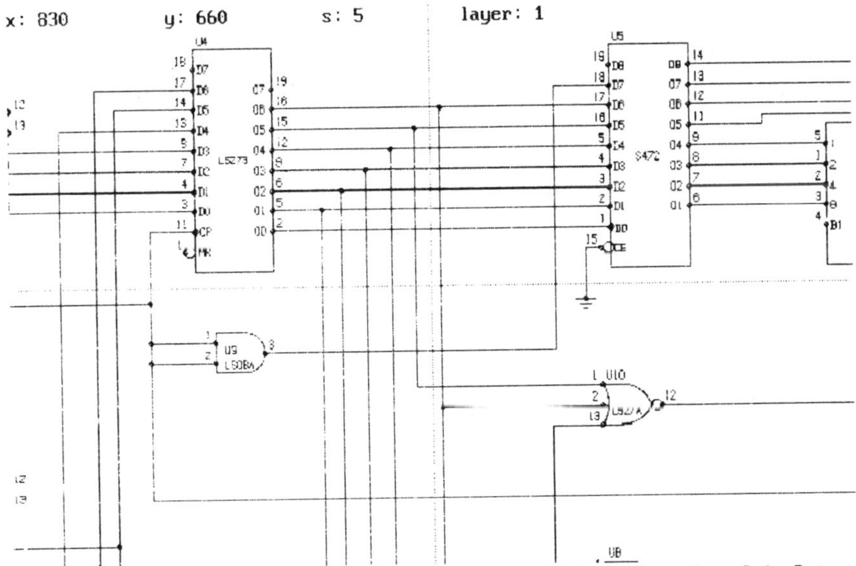

Figure 7-1 A *DC/CAD* schematic capture screen.

Part references are also added to the devices only after they are placed on the schematic. Devices with multiple gates may be assigned at the same time the part is named. Each device has two lines of text associated with it: part reference and part value. The size and location of the text is infinitely variable.

Lines and wires (treated separately) are started by moving the crosshairs to the beginning of the line and pressing <ENTER>. Thereafter, each click of the mouse changes the direction of the line or wire while maintaining connectivity. Pressing <ENTER> again ends the line. You can draw 45-degree diagonals while in the ortho mode, and buses are supported.

Schematic editing features are good. Parts can be moved, deleted, rotated, and copied as either single objects or as blocks. Find, locate, and jump commands are available too. There is an undo command that restores the last deletion.

Schematic netlists are generated outside the program using a DOS utility. The netlist format is proprietary and recognized only by *DC/CAD*'s PCB layout program. The BOM (bill of materials) netlist can include pricing and vendor sources for the components.

PCB Layout

The PCB layout mode is actuated manually after the program is started. The menu is identical for both schematic capture and PCB layout. Going into the PCB mode simply turns off some menu commands and activates others.

The program can read *DC/CAD* netlists only. However, you can buy a $195 software conversion option that translates many popular netlist formats into *DC/CAD* format for PCB layout, including *OrCAD* and *FutureNet*. But like most translators, it cannot convert device outlines, which means you have to go into the netlist and manually add the *DC/CAD* outlines yourself. Forward and backward annotation is provided between the *DC/CAD* schematic capture and PCB layout netlists.

The PCB library contains outlines for about 70 devices. There is a good sampling of DIP (dual inline), SIP (single inline), PGA (pin grid array), and SMD (surface mount device) packages, but many popular outlines are missing and resistor and capacitor outlines are limited to a single half-watt-sized device. But again, new outlines are easy enough to make using the schematic capture tools. No special library editor is needed.

Before you can place parts, you must define the circuit board outline. The maximum board size is 32 × 32 inches.

Parts are placed on the board by first extracting their outlines from the schematic capture netlist. An automatic parts placement routine is available. However, it is a rudimentary placement routine that needs a lot of manual cleanup. Parts can be glued in place prior to auto placement and forbidden zones where parts are not allowed can be defined.

Parts can be rotated or flipped to the other side of the board for SMD applications, but only after placement. A ratsnest is not available for parts placement but is available for routing.

The autorouter is a costed rip-up router. You tell the router what tracks are permitted and which are not, and it does its best to complete the job. Routing is gridless, which means the router adheres exactly to your clearance specifications; if you say 10 mils between tracks, 10 mils is what you get — every time.

Although the track width is limited to one size per session, you may specify which nets you want routed at the current width. For example, you could specify wider tracks for the power supply and ground paths on your first pass and follow up with a narrower track for the signal paths on your next pass. Up to 32 copper layers may be autorouted

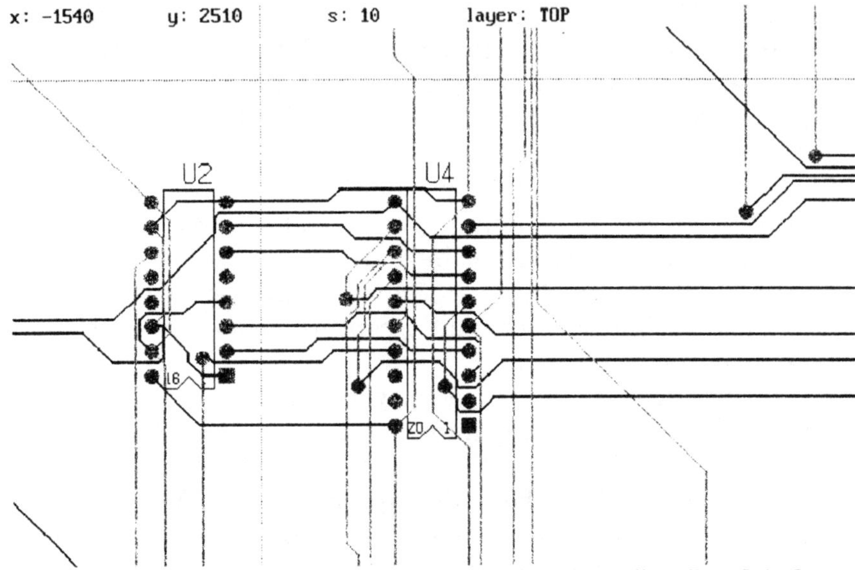

Figure 7-2 A *DC/CAD* PCB layout screen.

at the same time, and eight of those layers may be used for common planes or ground planes. All 64 drawing layers can be defined as copper.

Unfortunately, the rip-up router is much slower than the Lee routers found in most mid-priced PCB layout programs. What is worse, the autorouter is a DOS utility — which means you have to exit the program to run it. And while the autorouter allows for manual intervention, you must stop the router, go back into *DC/CAD* to make your changes, then return to DOS to continue with the autorouting sequence.

DC/CAD has a semi-automatic routing routine that finds the best path between two pads. Unfortunately, the

Table 7-1 Circuit Design Rating Summary

Schematic Capture:

Drawing	Editing	Library	Netlist Support	Ease of Use
good	excellent	good	poor	good

PCB Layout:

Place	Route	Library	Netlist Support	Ease of Use
good	good	good	good	good

gridless strategy of the rip-up router often has the tracks hug close to pads and other tracks, resulting in tracks that meander rather that taking a shorter, more direct route.

The solutions are to manually place the first few tracks yourself or edit them after the routers are finished. Fortunately, you do not have to delete a track to change it. *DC/CAD* allows you to move a track without breaking its connectivity.

Design rule checks are built into the program at each stage of operation — and they can sometimes be a real pain. For example, when compiling the netlist in DOS, the least infraction aborts the procedure and directs you back to the schematic capture program to correct it. The same is true for all PCB layout and ratsnest netlists used by *DC/CAD*. You often find yourself constantly switching between the program and DOS.

DC/CAD generates DXF, Gerber, and N/C drill files. All three output options are available from the *DC/CAD* main menu. Component, solder, and silkscreen masks are also available from the main menu.

Printer output is limited to a select few dot-matrix printers, notably Epson's FX80 and LQ-1000 printers, and is done through a sequence of events that has you create an intermediary file from *DC/CAD*'s main menu. A DOS utility is then used to convert the file into printer format for printing. Plotter support is broader, but mainly limited to Hewlett-Packard compatibles, and does not require a DOS conversion utility.

DC/CAD has everything you could ask for in a circuit design program: schematic capture, PCB layout, autoplacement, a rip-up autorouter, Gerber and N/C file output, and a superlow $495 price tag. However, the libraries are small, autorouting is slow, and printer support is limited. But for the serious hobbyist or fledgling business, it is a bargain that is hard to pass up.

Chapter

8

EE Designer III, Visionics Corp.

EE Designer III is a fully integrated circuit design package. In addition to schematic capture and PCB layout, you get automatic parts placement, an autorouter, and analog and digital circuit simulators that catch design errors before they become hardware problems. Best of all, the package costs only $995 — considerably less than some circuit design programs without schematic capture. However, the program is not easy to learn or use.

Although the program will run with 640K of RAM, any serious user should consider installing expanded RAM. Up to 8 MB of LIM 3.2 or LIM 4.0 is supported. A hard disk is required, and an included copy protection hardware connector (called a dongle) must be installed in the parallel printer port before the program will load.

Both the schematic capture and PCB layout programs are accessible from a common menu that forms a seamless link between the two programs. But learning and using *EE Designer* is not easy. First you have to memorize dozens of commands, then navigate a labyrinth of menus to find them. Fortunately, each command has a related keyboard key that reduces the number of times you have to drag and click on the mouse, making the program easier and faster to use — once you have mastered the technique. The automatic parts placement and autorouting routines also make using the program easier.

EE Designer's pan is manual and not automatic, having you move around the screen by selecting the pan command from a menu and indicating how far you wish to move, AutoCAD style. However, pan can be used during manual part and track placement without interrupting the placement procedure, as can zoom. Zoom range is 100 to 1 with a 5-to-1 magnifier that scopes in to fill the screen with a single schematic symbol or just three pads of a standard 14-pin DIP.

Although there is no help screen, there is a command description bar at the top of the screen that explains enough about the highlighted command to keep you from constantly running to the user's manual for help.

Schematic Capture

The program defaults to the schematic capture mode. The schematic-capture component library consists of about 1250 devices stored in a single library module that consumes almost 4 MB of hard disk space. Not only is the number of devices small for a program of this caliber, but you will also want to keep the 40-page library catalog close at hand because there is no on-screen listing for the devices.

Components are called up from the library by typing their name into a component reference box in one of the many menu paths. Multiple placement of a part can be made using a single mouse click. New parts are added by erasing the old part name from the component reference box and entering the library name of the new device. Components can be rotated and mirrored during or after placement. *EE Designer* also has an automatic parts placement routine that optimizes the parts location on the drawing after you have entered them, but the layout is not always the most practical for a schematic, and the function is really intended to benefit the PCB layout part of the program.

References to the device can be placed on the schematic at the time the part is laid down or after the drawing is completed. If you choose the automatic naming routine, the parts are numbered in the order they were placed on the drawing, with *EE Designer* grouping multiple-gate packages in a sequence that is beneficial to the PCB layout software.

Using the library editor, you can change, move, delete, or add reference lines to a component. There is no limit to the number of reference lines you can have, their location, or their font size — provided it is within the scope of *EE Designer*'s general editing skills. Parts modified in this way are not saved back to the component library, but are instead saved in *EE Designer*'s own database along with the drawing files associated with them.

Lines are started by clicking the mouse button once and moving the wire to its destination. Another click of the mouse button changes the direction of the wire. To end a wire you have to go into a menu or use its keyboard equivalent.

Editing features are excellent, provided you can put up with all the menu changes. Parts can be moved, deleted, changed, replaced, or duplicated on command as either single objects or as blocks. Unique to *EE Designer* is a locking function that prevents the editing of worksheet parts that are listed as locked in place.

Like all schematic capture programs, *EE Designer* uses a proprietary format for its netlists that is recognized only by *EE Designer* products. And while *EE Designer* netlist format is easily converted to other popular schematic capture

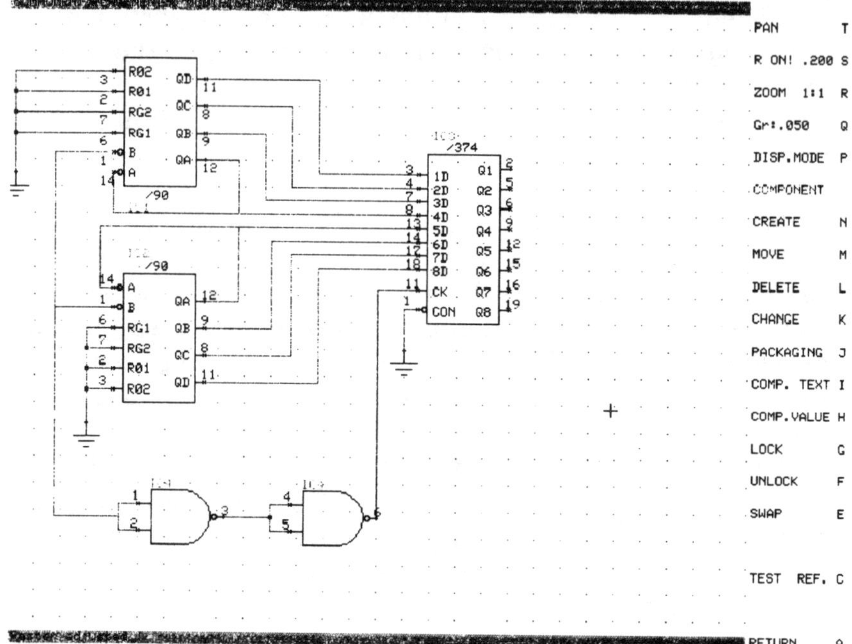

Figure 8-1 An *EE Designer III* schematic capture screen.

and PCB layout netlist formats, no such conversion utilities are included.

EE Designer has a most comprehensive BOM (bill of materials) netlist and includes pricing for individual components by vendor as well as a total price for the completed board.

PCB Layout

The PCB layout mode is selected from the common opening menu before the program is started. The screen is identical for both schematic capture and PCB layout, but the command menus are different.

Because the software is integrated, the PCB program accepts only *EE Designer* netlists; no format conversion utility is included. But for $195 extra you can buy the NESTIE option that converts *OrCAD* and *Schema* netlists into *EE Designer III* format. The schematic capture and PCB netlists are very tightly linked using both forward and back annotation. A change in either the circuit board or schematic prompts a user-approved update to the other's netlist so that the two are never different.

The 1250-device PCB component library is inextricably tied to the schematic capture component library, and any part that does not exist in the schematic capture library cannot be placed on the circuit board. For PCB layout, this intimacy is stifling because it does not allow you the freedom of specifying generic devices. Visionics says a larger library will be shipping soon.

Before you can extract device outlines from the schematic netlist, you must define a circuit board outline. Maximum board size is 32 × 32 inches.

EE Designer has an adequate automatic parts placement routine that determines a part's location by counting the number of attached wires and placing those with the higher closer together. Parts may be manually positioned and locked in place, forcing the placement program to work around them. Parts can rotated before or after placement and flipped to the other side of the board for SMD layouts. A ratsnest is available for placement, with the power supply lines changing connectivity from part to part as the device moves away from one component and closer to another.

Two Lee-based autorouters are included, both with fixed cost functions that cannot be changed. The first is a totally automatic router that routes the traces on a .025-inch grid. You simply enter the signal and power track widths, and the router takes it from there. Even via optimization, the removal of temporarily placed vias created during the initial routing pass is performed automatically. Unfortunately, the router is limited to just two layers: component side and solder side. But because of the tight .025-inch grid spacing, this router often achieves 100 percent completion — usually at the cost of a high via count. On both the automatic placement and our benchmark placement layouts, the router missed only one track.

The second is an interactive router on a .050-inch grid. This autorouter permits routing of up to 12 copper layers using a variety of routing routines that includes supply, signals, pin-to-pin, and windows-only tracks. The router may be programmed for nonstop routing or halted for intervention on a route failure. Like the above router, you may specify different track widths for power, ground, signal, and memory tracks. You can also designate protected zones where traces are prohibited. During initial routing, temporary vias are placed on the board, which may be removed using the optimize command. Although the completion rate of this router is less than the .025 router because of the larger grid, producing six failed tracks with automatic placement and seven failed traces on our benchmark layout, it permits greater user control (though not as much as if the cost functions were programmable).

Although the menu displays a rip-up router option, it is not available in the United States. But for $750 you can buy a *MaxRoute* (list price $6,000) rip-up router interface.

Either router requires manual handling of the following items. Neck down is done by adjusting the width of the trace on the fly during manual routing of the track. Curved tracks with a constantly variable radius, ortho tracks with 45-degree bends, and free tracks (no ortho restrictions) are also manually placed. A single track may contain any or all of the above-mentioned elements. Tracks are started by clicking the mouse button once and moving the track to its designation and clicking again to change direction or width. You can change layers while routing a track, with

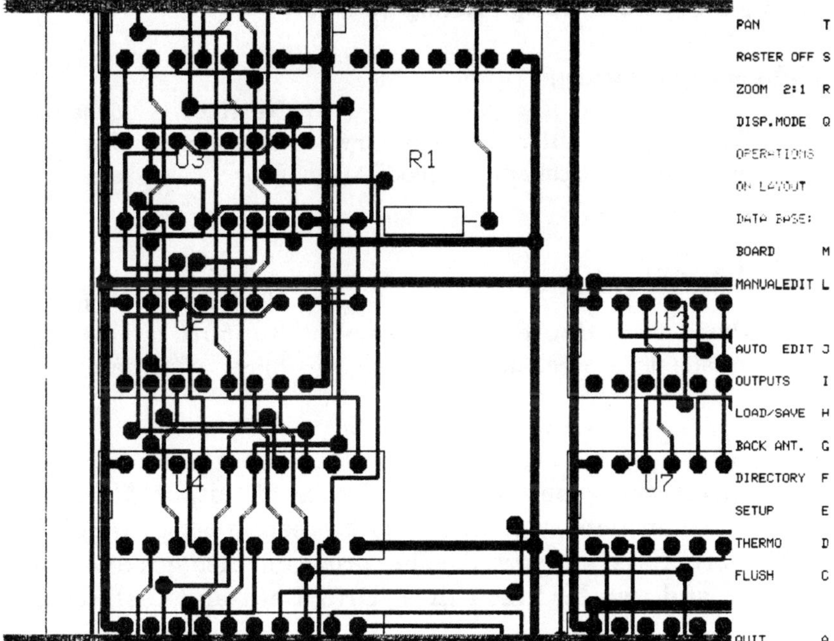

Figure 8-2 An *EE Designer III* PCB layout screen.

the program automatically inserting a via. Placed tracks can also be moved from one layer to another. Ground planes can be placed on any of the 12 layers and tied to any net, making it easy to create input guards for high-impedance analog circuits. Up to eight different ground planes are possible. Gate and pin swapping are supported, as is component renaming. Block functions include delete, move, and rotate for both parts and tracks.

The design check rule sections are thorough, but leave something to be desired. Although it lists unconnected nets and track violations, you have to toggle between a netlist display and the board layout to locate the problems. Too bad the two are not integrated in one screen.

Table 8-1 Circuit Design Rating Summary

Schematic Capture:

Drawing	Editing	Library	Netlist Support	Ease of Use
excellent	excellent	good	poor	fair

PCB Layout:

Place	Route	Library	Netlist Support	Ease of Use
excellent	excellent	poor	fair	fair

EE Designer generates a full range of output files, including component, solder, and silkscreen masks. *AutoCAD* DXF format is supported. The program also produces a Gerber and two N/C drill files (Drill Data and Drill Template).

A dot-matrix or laser printer may be used to produce a hardcopy of the artwork. The artwork can be sized to fit the page, scaled to a model size, or printed in actual size. However, the track width is not realistically reproduced, making printer outputs useful for proofing only. Pen plotters, on the other hand, draw perfect foil patterns. All printer and plotter adjustments, such as offset and pen velocity, can be set from the software.

EE Designer III is a very sophisticated, tightly integrated program that contains everything you need to go from design concept to actual printed circuit board in one, low-priced package. However, the small size of the components library is restricting, and it is not the easiest program to use. Still it's a lot better than many modular programs selling for a lot more, and a package any serious designer should consider.

Chapter

9

HiWIRE II, Wintek Corp.

HiWIRE II is a fully featured circuit design program that comes with both schematic capture and PCB layout software. However, the link between the schematic capture and PCB programs is tenuous at best, and using *HiWIRE* is like working with two unrelated packages. The program is easy to learn, moderately difficult to use, and short on features. List price is a low $995.

Before we proceed with the review, it is only fair to warn you that the *HiWIRE II* software we received for review is a beta copy. And while Wintek assures us that this is close to what to expect, some particulars may change between now and its release.

Although the program itself does not support expanded memory, the optional PCB autorouter can use up to 15 MB of extended memory or 32 MB of expanded LIM memory. A hard disk is required and the hardware protection dongle that accompanies the package must be installed for the program to work.

Both the schematic capture and PCB software are available from a common screen. Both programs use identical menu formats and commands so that you only have to learn the routine once. But tersely labeled commands and a quirky menu selection routine make learning and using the program more difficult than it should be.

HiWIRE II has a scrolling auto pan that is in effect during both schematic and PCB line placement — quite an improvement over *HiWIRE Plus*. However, you still have to escape the placement routine to zoom in or out. Zoom range is 128 to 1, with the end result being an E-size drawing that is about the size of a postage stamp on the screen; in the PCB layout mode, the highest magnification fills the screen with an 8-pin DIP. The advantage is that you can have more than one worksheet on the screen at the same time. *HiWIRE* also has a windows function that lets you display two different areas of the screen alongside each other, but for reference only because only the current screen is active. There is no help screen or menu.

Schematic Capture

The schematic capture component library comes with 850 devices distributed among seven library modules. Most of the part names are generic, like PNP or opamp, except for the TTL chips which have a 74xxx designation. Unfortunately, only one library can be active at a time, which means you have to delete the current library and load a new library when changing device types.

Components are called up from the library by clicking on a menu command and typing in the device name at the prompt. A screen listing of the library devices is available, but only for reference. Devices cannot be selected from the screen menu, which is unfortunate because many device names include case-sensitive characters (e.g., "formC" is not the same as "formc"). Every part placed on the worksheet must be typed in individually because *HiWIRE* does

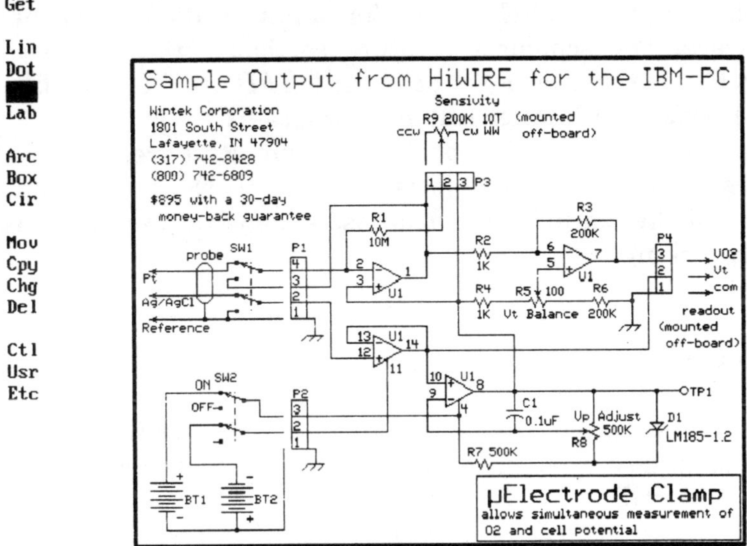

Figure 9-1 A *HiWIRE II* schematic capture screen.

not remember the last part entered or have a multiple placement routine.

Parts can be rotated or mirrored, but only after they are on the worksheet. However, certain combinations of the positioning edits cause other body parts to overlap and mask one or both references lines — neither of which can be moved. Parts also have no schematic reference until manually assigned after placement using an edit command.

Only ortho wires or lines are permitted, no diagonals. Once a wire or line is started it can be continued indefinitely, making turns with each successive click of the mouse until ended.

Among the editing options are copy, move, delete, and find. Block support, which *HiWIRE* calls BIND, includes all normal editing and drawing features.

Included is a netlist converter that converts schematics created with *FutureNet, OrCAD, Schema,* and *Tango* netlists into *HiWIRE* format for input to its PCB layout software. The schematic capture section can generate either PostScript LaserWriter or *AutoCAD* DXF output files for interface with other software programs.

Despite the inclusion of PCB layout software, *HiWIRE* offers very little value. It is difficult to work with, and many of the features found in less expensive programs are sadly lacking.

PCB Layout

For an integrated circuit design package, *HiWIRE II* has an amazingly weak interface between the schematic capture and PCB software. It is more like what you would expect from a modular package, with the pin connections being the only thing you can extract from the schematic. The advantage is that *HiWIRE* can and does convert a wide assortment of schematic capture netlists into *HiWIRE* format (18 in all, with 7 of those different *HiWIRE* formats) because so little information is needed for conversion. The downside is that you have to do most of the PCB layout work by hand.

The PCB library contains about 140 patterns that are an array of pads spaced to accommodate a specific device footprint. There is also a separate library of pad sizes and shapes that can be used to create or modify PCB patterns, and another library of silkscreen outlines.

All board layout is done by hand, even if you intend to import a schematic netlist later. Parts are called from the library by clicking on the menu command and typing in the device name at the prompt. A screen listing of the library devices is available, which is a godsend because the device names are case sensitive (dip16 is not the same as DIP16), and being able to point and click on the desired object saves a lot of time and frustration. Devices are

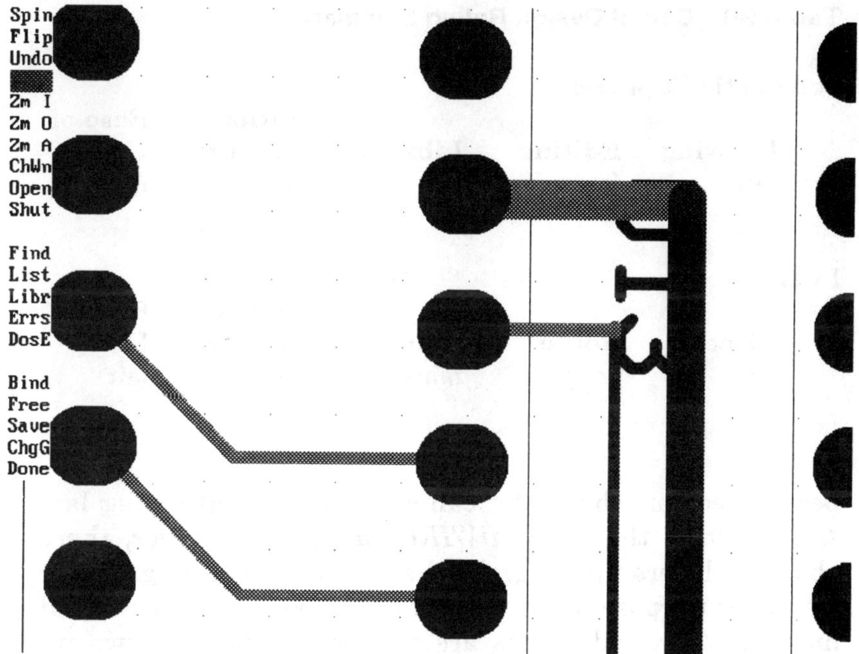

Figure 9-2 A *HiWIRE II* PCB layout screen.

automatically numbered in sequence as they are placed on the circuit board. Parts can be rotated or flipped to the opposite side for SMD layout only after they are placed. Every part placed on the board must be typed in individually because *HiWIRE* does not remember the last part entered or have a multiple placement routine. The ratsnest is available only after all parts in the netlist are placed on the board. Maximum board size is 64 × 64 inches.

The optional autorouter ($1,295 for the regular version and $1,695 for the EMS version) is Lee-based and allows the user to program four costing functions. Among the choices are the number of routable layers, the number of routing passes to be made, and the number of allowed vias

Table 9-1 Circuit Design Rating Summary

Schematic Capture:

Drawing	Editing	Library	Netlist Support	Ease of Use
fair	fair	fair	fair	fair

PCB Layout:

Place	Route	Library	Netlist Support	Ease of Use
poor	good	fair	good	fair

per connection. The router can route up to eight copper layers, which is the most *HiWIRE* supports. However, there about 60 layers up for grabs that the user can program for whatever purpose, for a total of 64 copper layers. An equal number of ground planes are supported, with the two innermost layers dedicated to power and ground, and they may be connected to any net. There is no upper limit on either the number of router passes or vias. Protected areas are supported. No benchmark testing of the autorouter was done.

Manual routing is done by choosing the line command from the menu and scooting the rodent along to its destination and clicking once to change direction or twice to quit. Track width cannot be done on the fly. You can change layers during routing, with automatic via placement. All copper layers may have a ground plane, and the plane may be connected to any net.

Design rule checks include unconnected nets, air gap violations, and missing footprints.

HiWIRE II produces Gerber, N/C drill, PostScript, and DXF files. Also generated are solder, SMD paste, and silkscreen masks. Although there is support for a wide range of printers and plotters, the program has virtually no control over the file output to a dot-matrix printer, which results in an A-size drawing using less than one-fifth the area of an 8½ × 11-inch sheet, and only slightly better control over a laser printer. Schematics produced by either are of very poor quality and are just barely usable as drafting proofs. If you are dealing with a plotter, however, you have absolute control over the instrument at all times.

HiWIRE II is a fairly difficult program to learn and use, but it contains all the elements needed for creating very complex circuit boards. Its support of 64 copper layers on a board size of 64 × 64 inches is unsurpassed. If you do not need these features, there are several easier-to-use programs to choose from.

Chapter

10

OrCAD, OrCAD Systems

OrCAD has two programs for circuit design use on the PC. The first is *OrCAD/STD III,* an excllent schematic capture package that is one of the industry's hottest sellers. Second is *OrCAD/PCB II,* a companion PCB layout program. There is a tight link between the two programs, with both forward and back annotation.

However, each program stands on its own, and they are run from DOS separately. Those who wish to have both *OrCAD/STD III* and *OrCAD/PCB II* programs accessible from a common menu can buy a third-party program called *Intelligent Menus Systems* from Velotec, Inc. of Anaheim, CA. Reportedly, *OrCAD/STD IV* will have an interactive menu.

Either program can run on a dual floppy disk system with 640K of RAM, but for serious work a hard disk is required. Expanded RAM and a math coprocessor are not needed or supported. A hardware protection key (dongle) is not required, and neither program is copy protected.

Figure 10-1 You can use the program *Intelligent Menus System* from Velotec to access *OrCAD/STD III, OrCAD/PCB II,* and the OrCAD utilities from a common menu.

OrCAD/STD III

OrCAD/SDT III was among the first schematic capture packages to appear on the scene and is still one of the most versatile and complete schematic capture programs around. It is easy to learn and use, comes with a huge component library, and carries a modest $495 price tag.

The program is moderately easy to learn and easier to use. Everything you need is in pull-down menus. *OrCAD/STD III* has autopan, but instead of scrolling smoothly, the screen abruptly jumps to the next area of the drawing indicated by the cursor movement — making part and wire alignment difficult on large drawings. Turning on the grid and zooming out makes placement easier, but you'll have to zoom back in to fill in the details. There is also a jump feature that lets you tag up to eight spots on the worksheet and jump to a spot by simply calling its name.

There is a macro function that lets you record often-used drawing routines, such as creating memory arrays or labeling a worksheet, then replay the routine via a single keystroke from the keyboard. Macros are created by turning on the macro recorder and letting it run while you make your moves. Hitting "M" turns the recorder off and saves the macro to file. Over 100 macros can be stored, provided their aggregate total does not exceed 64K.

OrCAD/STD III's component library is made up of 11 library modules. Included among the 3500-plus components are several thousand logic chips, 100 Intel and Motorola processors, 40 PAL chips, and a comprehensive assortment of RF and microwave devices.

Components can be called from the library by entering its generic name from the keyboard or by clicking on the part from a scrolling library menu. The selected component can be placed on the screen as many times as you wish by simply double clicking on the mouse. Parts can be rotated and mirrored during or after placement.

Part reference annotation is done from an external utility, not from the drawing program. Device references are assigned in the order in which the parts were placed on the worksheet. Multiple packages are grouped in sequence beginning with the first object drawn of that type. Objects that have manually edited references are assigned a new reference, which have to be changed back following annotation if you wish to keep them. There are eight user-defined reference lines for each part, and each line can be moved freely and independently about the screen. Any or all of the reference lines can be hidden.

Wires are placed on the schematic using a six-step menu sequence. However, once a line is started, only two mouse

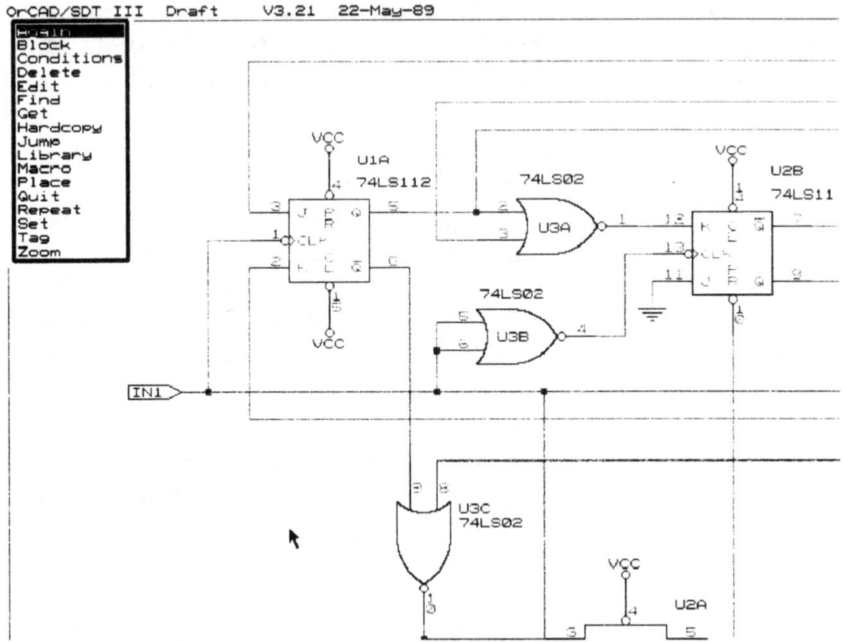

Figure 10-2 An *OrCAD/STD III* schematic capture screen.

clicks are needed to continue its path and a macro can be used to reduce the number of start-up steps.

Most of OrCAD's editing commands are done via blocks. To move an object, for example, you first define a block that includes the desired object, then move the block. Unfortunately, everything else in the block also moves along with your object, and you may have to delete unwanted items after you place the block. The only edit command that is object specific is delete.

Although OrCAD would like you to link *OrCAD/STD III* to their *OrCAD/PCB II* PCB layout package, they realize many users use this easy-to-use schematic capture program as a front end for a variety of different engineering

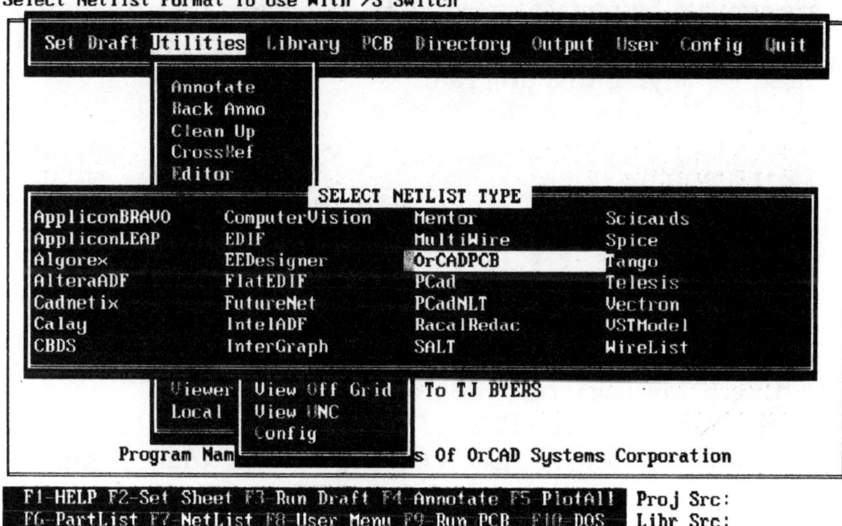

Figure 10-3 OrCAD supports 28 different netlist formats.

programs. Among the 28 output formats supported are SPICE, *FutureNet, Tango,* and *EE Designer.*

OrCAD/STD III has two print options, one from the screen and the other from an external utility. Copies made on a dot-matrix or laser printer from the screen option are distorted (with circles that look like eggs), whereas hardcopy made from the external utility is of excellent quality. Unfortunately, you have no control over a printer or plotter from the utility; all page and drawing adjustments have to be made in the printer or plotter itself.

OrCAD/SDT III has a design check utility that removes duplicate wires, buses, and junctions, and it displays warning messages advising of other duplicate objects.

OrCAD/STD III is undoubtedly one of the best schematic capture programs around. The drawing and editing functions are a pleasure to use, and the netlist format is recognized by almost every PCB layout and circuit simulation program in the industry. You certainly will not get fired for buying this program.

OrCAD/PCB II

OrCAD/PCB II is a fully featured PCB layout program that contains nearly everything you need for PCB design, including a rip-up router, yet it lists for a moderate $1,495. The program is easy to learn and use and has a tight link with schematic capture program.

Although *OrCAD/PCB II* does not support expanded memory, the new release, *OrCAD/PCB III* (due in March 1991), will have support for up to 4 MB of LIM EMS.

The fact that you can extract device outlines from the schematic netlist makes the program very easy to use. Everything you need is in pull-down menus, and once the router starts, the program is on autopilot.

OrCAD's netlist is undoubtedly the most popular in the industry and is supported by most schematic capture programs in one form or another. However, you will probably end up doing a lot of netlist tweaking either inside the schematic capture program or on the netlist file itself using an ASCII editor because most programs cannot or do not include the OrCAD device outline as part of the netlist. The link between *OrCAD/STD III* and *OrCAD/PCB II* is very tight with both forward and back annotation.

The library contains outlines for over 300 devices that include a generous assortment of connectors and SMDs. A board outline, which can measure up to 32 × 32 inches, is not required before parts can be placed. But you do have to define a working area before the device outlines can be extracted from the schematic netlist.

However, you are not confined to the work area for placement. It is only required to load the parts, and you have the entire 1024 square inches to work from while manually moving the parts to their desired locations. Parts may be rotated and flipped to the solder side for SMD placement during and after parts placement. Both a ratsnest and a force vector display are available for placement. *OrCAD/PCB III* is said to have an automatic placement routine.

The Lee-based autorouter has six user-programmable costed functions that the program calls strategies. The strategies range from defining when a 45-degree track is permitted to no vias allowed. However, only one strategy and two layers can be in effect at a time. If you want to route four layers, you will have to do two autoroutes. If you are still left with unconnected tracks, you can change the strategy and try again — or use the rip-up router to place those dangling ends. You can define protected zones where tracks are prohibited. On our benchmark layout using the extensive option on OrCAD's fixed 5-mil grid, the autorouter scored 100 percent completion.

Autorouting track width is limited to one size per layer; up to 16 layers are supported. To route tracks of different widths on the same layer, you have to autoroute them one net at a time or lay them down manually.

Manually routing tracks involves a rather lengthy sequence of events that has you walking through six menus. However, once a track is started, only two mouse clicks are needed to continue its path, and the width of the track may be changed on the fly for necking down. Via placement may be automatic or manual, as you wish. You can also move a track that is in place while maintaining the connectivity of the track. All of the 16 layers can have a ground plane, and the planes may be connected to any net.

Panning is automatic, but it does not scroll smoothly. The screen abruptly jumps to the next area of the drawing indicated by the cursor movement. Zoom range is 100 to 1, with the highest magnification displaying about half a

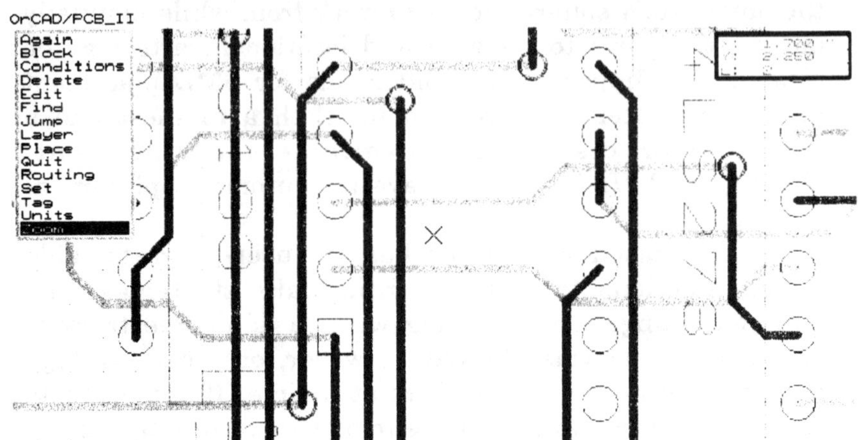

Figure 10-4 An *OrCAD/PCB II* PCB layout screen.

14-pin DIP chip. Both pan and zoom are available during parts and track placement. Block functions include move, copy, delete, save, and get. Find, locate, and jump commands are also supported; a help screen is not.

The design rule check logs air gap violations in a disk file, citing the coordinates of the violation that you will have to manually compare to the screen. The only other design check reporting occurs when doing an autoroute, in which the number of nets, vias, and unconnected nets are displayed at the bottom of the screen.

OrCAD/PCB II generates both Gerber and N/C drill files. Other output files include solder and silkscreen masks.

Either a dot-matrix or laser printer can be used to produce draft or final artwork in a range of scaled sizes. With some fine tuning on track width and spacing, a dot-matrix hardcopy of 1X artwork is suitable for prototype work and hobby projects. Of course, a Hewlett-Packard or Houston Instrument plotter drawing, both of which are supported,

Table 10-1 Circuit Design Rating Summary

Schematic Capture: OrCAD/STD III

Drawing	Editing	Library	Netlist Support	Ease of Use
good	good	excellent	excellent	good

PCB Layout: OrCAD/PCB II

Place	Route	Library	Netlist Support	Ease of Use
fair	good	excellent	excellent	good

does a better job. But PostScript users, which *OrCAD/PCB II* also supports, may dispute that point.

OrCAD/PCB II is an excellent layout program that has everything the professional PCB designer needs, and the price is affordable. However, it lacks the sophistication you have come to expect from OrCAD — something OrCAD promises to correct in *OrCAD/PCB III.*

Chapter

11

ISIS Supersketch and PCB II, Labcenter Electronics

Labcenter Electronics serves up a two-package circuit design deal that is aimed squarely at the performance- and budget-conscious hobbyist. Both the *ISIS Supersketch* schematic capture and *PCB II* printed circuit board layout programs are easy to learn and use, support several drawing and editing features, work with either floppy or hard disks, and sell for a low $149 each.

However, each program stands on its own; they are not interconnected. *Supersketch* cannot produce netlists and *PCB II* does not read them. But Labcenter has an industrial line of fully interactive schematic capture and PCB layout programs, called *ARES,* that are compatible with *Supersketch* and *PCB II* files should you decide to upgrade. Prices start at $599.

Either program can run on a single floppy disk with just 512K of RAM, but most users would opt for two floppies or a hard disk. Expanded RAM and a math coprocessor are not needed or used. A hardware protection key (dongle) is not required, and neither program is copy protected.

Commands are invoked from a combination of pull-down menus, dialog boxes, icons, and keyboard strokes — an arrangement that is familiar to any Macintosh or *Windows* user. As such, the programs are easy to learn and easy to use. A help screen is not provided, nor did we find one necessary. However, the use of European symbols and terminology takes some getting use to.

Two methods of panning are provided. With no place command in effect, you pan by clicking on the overview display found in the upper right-hand corner. This moves the working window around to all areas of the drawing. However, autopan is activated when placing a component, trace, or wire, causing the screen to jump by about one-half of its view when the cursor bumps against a screen edge. Zoom range for *Supersketch* is 20 to 1, while the zoom range for *PCB II* is 50 to 1. Zoom cannot be used when placing a part or wire. Neither program supports macros.

Schematic Capture

Supersketch ($149) is an entry-level product that lets you draw schematic diagrams on pages up to A0 in size (the European equivalent of an E-size drawing). The program comes with a single device library that contains 58 generic devices, such as transistors, op amps, and connectors.

Supersketch Plus ($199) is identical to *Supersketch* with the addition of five more libraries that bring the total component count to over 3300. The new libraries serve up a healthy portion of logic, analog, and memory devices. But the bulk of the component count is in TTL chips with different prefixes, like 74LS00 and 74ALS00. A count of different device types puts the number at 600 — enough for most drawing chores.

New library parts can either be made from scratch or by modifying an existing part using *Supersketch*'s library editor. The editor is available from a pull-down menu, which means you do not have to interrupt the drawing process to create a new device.

There is an on-screen listing of the library's contents. You extract parts from the libraries by highlighting the desired part from the list and clicking on it. This transfers the component to a device window, where it is stored. After you have finished selecting the parts, you close the library window.

At this point no devices are on the drawing. You now go into the device window and select the part you wish to place. As long as a device remains highlighted, it may be placed as many times as you wish by simply clicking on the mouse. The components in the device window become a permanent part of the drawing's database file and need to be loaded only once.

Commonly used symbols, like input/output terminals and ground, are not included in the device window, but are instead available from the tool kit window as icons. Clicking on any of these dozen or so devices makes them available for placement.

Components may be rotated in 90-degree increments or mirrored right to left before or after placement — but not during the actual placement. References to the device are done after the part is in place. Automatic numbering of reference numbers is available, but you can only do one component type at a time (resistors or ICs, for example), and the program does not check for duplicate entries. Most annotation is done by hand using the device editor to change, move, or delete reference lines. There is no limit to the size of the font or the reference's location on the drawing.

To draw a wire, all you do is click once on its origin and once on its destination. *Supersketch* has a wire autorouter that draws the line for you, automatically selecting the

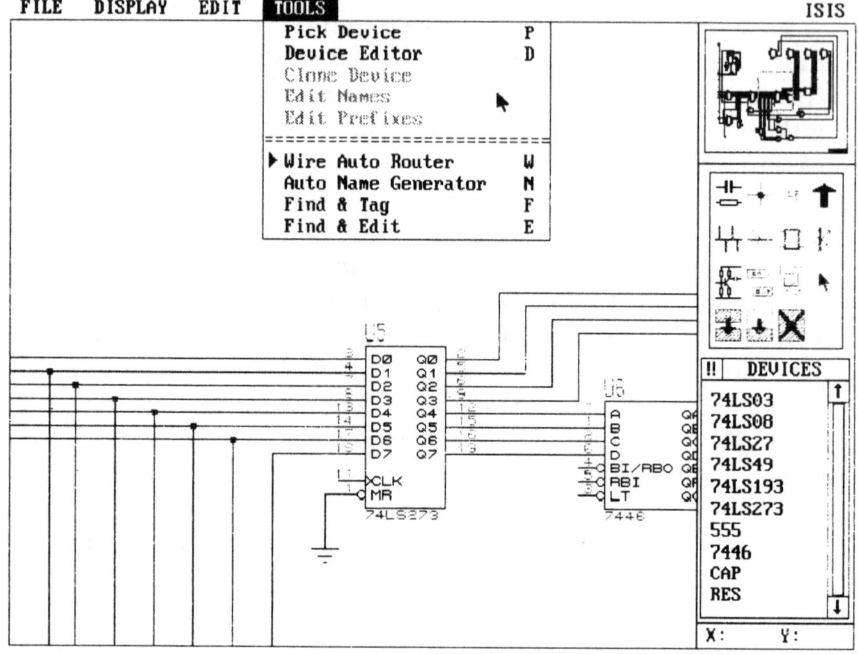

Figure 11-1 A *Supersketch Plus* schematic capture screen.

best route and adding whatever corners are needed to make it ortho. Lines can also be drawn manually by clicking the mouse button once for each time you wish to change direction, including drawing diagonals at any angle. The line automatically terminates and is placed when you click on a component terminal or another wire; no dangling or unterminated wires are permitted. Buses are supported.

Editing commands let you move, delete, rotate, mirror, and copy symbols. When a part is moved, the connected wires rubberband (expand or contract to its new position). Once the part is placed, the autorouter automatically cleans up the rubberband lines into short, ortho lines. Lines can be moved manually too. Wires and symbols can be locked in place to prevent their editing. Parts defined as a block may be moved, deleted, rotated, copied, and saved. The undo command sequentially replaces all deleted parts with each use, back to square one if needed. Jump and find are supported, but not locate.

Supersketch supports a wide array of printers and plotters, including many popular dot-matrix and PostScript laser printers. The schematic can be printed in portrait or landscape orientation, and the 4-to-1 scaling factor lets you print an A0-size drawing on an 8½ × 11-inch page.

For drawing schematics, *Supersketch Plus* is unquestionably one of the best. It is easy to use, has a good component library, abounds with drawing and editing features, and supports a wide range of print and plot options. But the lack of netlist output is a serious drawback, and one that is corrected only by spending a lot more money.

PCB Layout

PCB II is Labcenter's low-end PCB layout package. The program is closely related to the higher-level *ARES PCB* layout package ($599), except that *PCB II* is totally manually operated. The ease of use is on par with *Supersketch*, as are the editing features, but the placing of parts and tracks requires more work than drawing a *Supersketch* schematic.

The program comes with a library of 70 component outlines that include several DIP and PLCC footprints. Included too are footprints for a number of relays, switches, and several different size resistors and capacitors.

Not included in the library are pads or vias. These are found in the tool kit window along with the track widths. Like the ground symbol of *Supersketch*, these are commonly used items that can be placed by clicking on an icon. The program comes with a list of 20 common pad sizes and

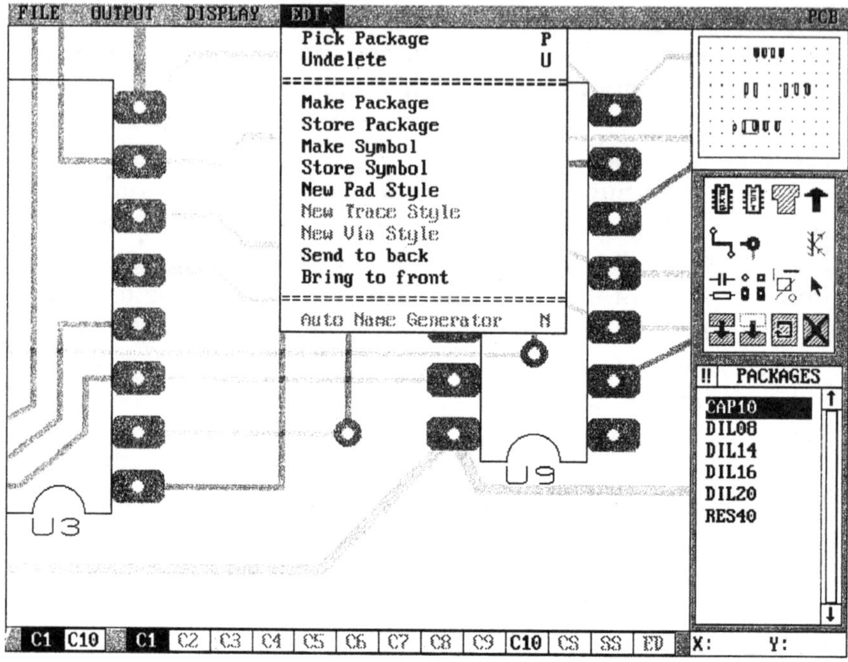

Figure 11-2 A *Supersketch PCB* layout screen.

patterns, but you may add as many of your own design as you wish.

Like *Supersketch*, component outlines are called from the library using a parts list. Again the selected outlines are stored in a component window, which is then used to place the outlines on the board. As long as the device in the component window remains highlighted, it can be placed on the board as many times as you wish.

Because *PCB II* does not read netlist files, each design is started from scratch by hand. However, subcircuits like memory arrays can be saved in a block file and used again in any number of layouts. Parts may be rotated in 90-degree increments or mirrored for SMD use before or after

Table 11-1 Circuit Design Rating Summary

Schematic Capture: Supersketch Plus

Drawing	Editing	Library	Netlist Support	Ease of Use
excellent	excellent	good	none	excellent

PCB Layout: PCB II

Place	Route	Library	Netlist Support	Ease of Use
good	poor	excellent	none	excellent

placement, but not during placement. A board outline is not needed before placing parts. Maximum board size is 30 30 inches.

Unlike *Supersketch*'s wire autorouter, *PCB II* tracks have to be laid down manually from start to finish. Tracks may be placed at any angle and any width, and via placement is automatic when changing from one layer to the other. However, the board is limited to just two layers of copper: top and bottom. Grid snap is incrementally variable between 5 and 100 mils, and can be turned off for near-gridless routing. Copper fill for current guards and ground planes is supported.

Devices that are moved after tracks have been attached drag their tracks along with them. But unlike *Supersketch,* you have to do the cleanup work yourself. Track routing and width are changed by laying down a new track over the old. *PCB II* recognizes it as a short circuit and removes the old track, leaving only your new path. You can remove vias and neck down using the same procedure. Find and search are supported.

There are no design rule checks, and all mistakes simply pass through to the final board product.

PCB II generates Gerber, N/C drill, and PostScript files, as well as silkscreen and solder masks. *PCB II*'s printer and plotter interface are identical to *Supersketch*'s, which means you can produce final artwork at scaled sizes (up to 4:1) on dot-matrix or laser printers and several pen plotters.

PCB II has a lot of nice features, including a large library of device outlines, an unlimited number of pad shapes and track widths, Gerber and N/C drill file output, and a low $149 price tag. But its lack of netlist input and limit of two copper layers makes it suitable only for occasional projects with few parts. For $450 more, you can buy the *ARES PCB* package that does all these things for you.

Chapter

12

SuperCAD, PCBoards, and PCRoute

This chapter reviews a trio of software programs sold by three different vendors that together form a complete circuit design package. The programs included are *SuperCAD,* a schematic capture program from Mental Automation; *PCBoards,* a PCB layout program from PCBoards; and *PCRoute,* an autorouter from RGH Software Design. All three are designed to stand alone or interface via common netlists to form an integrated circuit design package. Each sells for $99, making the combination a very cheap complete design package. Each of the programs is relatively easy to learn and use.

SuperCAD from Mental Automation

At $99, *SuperCAD* is the lowest-priced schematic capture program in this book. But don't let the low price fool you. This package has a full complement of drawing and editing features and is easy to use. The only thing it lacks is a decent component library.

It can run on a dual-floppy system, but prefers a hard disk. Expanded memory and a math coprocessor are not needed or supported. A hardware protection key (dongle) is not needed, and the software is not copy protected.

The component library comes with only 350 devices, 233 of which are TTL logic chips, distributed among 11 library modules that are always active. There is a good variety of device types, including DSP (digital signal processing), PLD, and microprocessor chips, plus DeMorgan equivalents. There just are not enough of them.

Components are called up via a window that lets you select the library module of your choice using the mouse. Once inside the library, you simply click on the part you want. Up to six devices may be moved from the window to an on-screen quick-access drawing cell that lets you place a part without having to go through the window routine again, making multiple placements a breeze.

Parts can be rotated, but the rotation point is outside the device outline, forcing you to move the part after rotation. Each worksheet also has four drawing planes that may be displayed individually or superimposed. CAD programs like *AutoCAD* often use layers to separate colors, but layering is rare among schematic capture programs. The best use of layering in schematic capture applications is to use the layers to separate the design functions on the worksheet; i.e., the clock chips are on one layer, the output drivers on another, and so forth. That way you can do massive editing of one layer without having to change things on the other three.

Components are placed on the worksheet with no references. References are assigned in a separate step in the order in which the parts were placed on the worksheet. However, each type of device is handled separately, with resister annotation a separate step from IC or capacitor

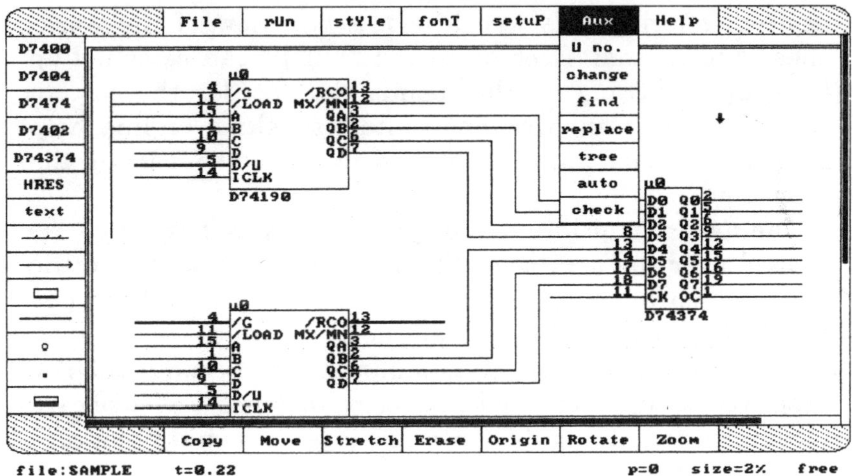

Figure 12-1 A *SuperCAD* schematic capture screen.

assignments. Each part has only one reference line, which is rigidly fixed for both placement and font size.

Wires are placed on the drawing by selecting the wire type from the drawing cells and clicking once to start the wire and again to stop the wire. However, each wire is a single entity, which means you cannot have continuous runs. You must end one wire, then begin the next at the termination point for continuous runs.

There is no autopan. Instead, scrolling is done by using the scroll bars at the right side and bottom of the screen or by using the keyboard's page up, page down, and arrow keys. In fact, keys play an important part in drawing and editing of the schematic, and in several cases there is no mouse equivalent of a keystroke.

The editing features are excellent, with move, copy, delete, change, replace, and find available for both single objects and blocks. There is also a design checker that points out drawing mistakes. Zoom, however, leaves a lot to be desired, offering only two levels: a full view of the worksheet and normal. Each of these functions can be called up by simply clicking on the menus that border the screen. *SuperCAD* has an excellent context-sensitive on-line help screen that tells you all you need to know about using the program.

The hardcopy printout is of good quality, but you have no control over the printer or the size of the schematic on the page, which means you may end up taping more than one schematic together. Plotters are not supported.

For its price, *SuperCAD* is amazingly well endowed. The program is fully featured and easy to learn and use. We recommended it highly for entry-level schematic capture work.

PCBoards from PCBoards

PCBoards is a manual PCB layout program that sells for an astonishingly low $99. But as its price suggests, it has limitations that make it applicable only for simple designs, like making breadboards or hobby projects.

PCBoards' hardware requirements are a miserly 384K of RAM and one floppy drive. Expanded memory and a math coprocessor are not needed or supported. Again, there is no hardware protection key, and the software is not copy protected.

PCBoards is easy enough to learn, but it may take some getting use to the sometimes awkward placement sequences. Nothing about the program is automated, forcing the user to essentially translate a schematic drawing into a PCB layout by hand. But that's a whole lot better than using transfer tape and a knife.

The component library, as it is, consists of DIP (dual in-line package) and SIP (single in-line package) pad patterns with standard 1/10-inch spacing. For DIPs, the footprint is selectable from 6 to 48 pins; SIP outlines range from 2 to 40 pins. Individual pads are used to place resistors and other devices.

Part placement consists of calling either the DIP, SIP, or pad command from the menu and positioning the cursor on the grid point that relates to pin number one of the device and clicking the mouse button or pressing a function key. Both DIP and SIP patterns may be rotated during placement, but it may take awhile to get the hang of this because the DIP pattern does not become visible until after you click, leaving you to guess about the outcome. Maximum board size is 6 × 13 inches.

Placement editing is very limited. Everything must be done as a block function. For example, to move a DIP part, you first have to define a block area around the pattern, erase it, then move the upper-right-hand icon to its new location and request a block place. Again, many placement surprises may result until you get the lay of the land because the outlines are not entities, but individual pads on the screen. Missing a row of pads when marking a block not only strips the pattern of needed pads, but leaves the forgotten pads behind to be reckoned with in a separate step.

Track placement is very much the same, with the user clicking on the beginning and then the end of the track, with no visual display until the track is in place. Only ortho lines are permitted, with short (one grid-dot) 45-degree bends available for memory routing. *PCBoards* only supports a top and bottom copper layer, and vias must be manually placed when changing from one to the other.

There are only two track widths: 20 mils and 50 mils. All tracks must first be laid down using the 20-mil format. Tracks that need to be fatter are then widened using the THICK command. Thick lines cannot be initially placed.

If the thought of manually routing a circuit board does not thrill you, you can have the *PCRoute* autorouter do it

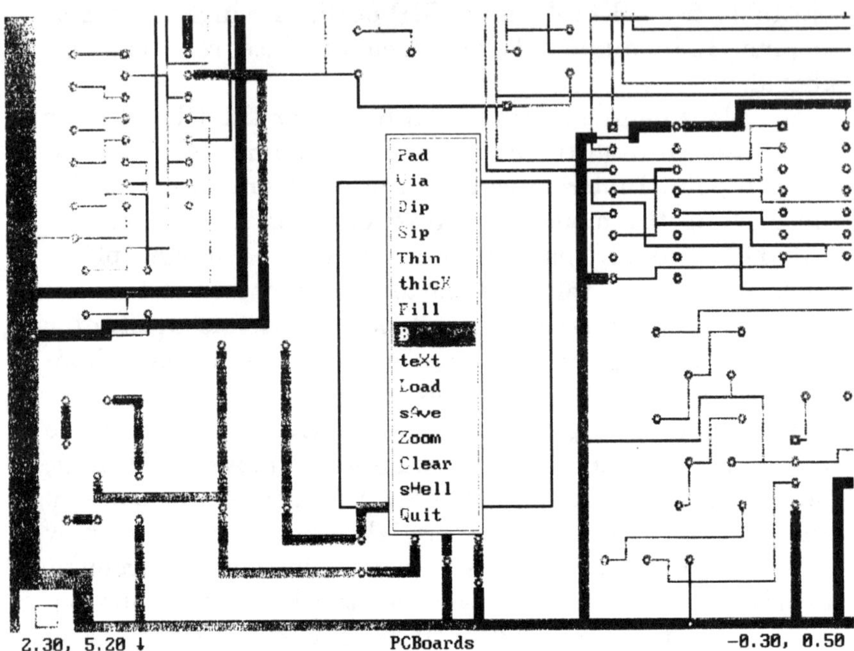

Figure 12-2 A *PCBoards PCB* layout screen.

for you by simply saving the layout to file and running *PCRoute*. *PCRoute* then combines the netlists produced by *SuperCAD* and *PCBoards* to route the board. However, the same trace width restrictions remain, and the schematic device outline interpretations are less than ideal. Which means you will probably end up doing a lot of manual touch-up in the end. No testing of our benchmark layout was done because of the track width requirements we built into the benchmark circuit that neither program can cope with.

PCBoards has a help screen, but no design rule check. Autopan is supported for all operations except track placement, where it is needed the most. Zoom has but two

Table 12-1 Circuit Design Rating Summary

Schematic Capture: SuperCAD

Drawing	Editing	Library	Netlist Support	Ease of Use
good	good	fair	fair	good

PCB Layout: PCBoards

Place	Route	Library	Netlist Support	Ease of Use
poor	fair	poor	poor	fair

levels, big and small, neither of which are practical for serious PCB work. However, a new zoom command with a wider range is in the works and should be shipping sometime soon.

There is no Gerber file generator, but there is a drill list of sorts. The drill list printout lists the coordinates of the pad and via holes and gives you a line in which you can manually write in the hole size on the hardcopy — enough information for a PCB service to manually program an N/C drilling machine. Solder and component masks are not part of the bargain.

Final artwork can be printed using either a dot-matrix or laser printer in 1X and 2X scale. For crude layouts with 50-mil tracks, the 1X dot-matrix pattern can be used to etch a circuit board. A better plan is to print the pattern in 2X scale and photographically reduce it — or use a laser printer. In the 1X mode, the laser printer can produce very good artwork on a clear sheet that is suitable for immediate use.

Support for Houston Instrument and Hewlett-Packard plotters will cost you $49 extra. The plotter software incorporates advanced features like pad shaving to maintain proper air gap spacing and fillets to relieve thermal stress at track corners.

PCBoards is not the greatest PCB layout program, but its super-low price makes it ideal for the hobbyist. However, the software is in a constant state of upgrade, with some of the shortcomings mentioned here scheduled to be corrected by the time you read this, and the upgrades are free for one year after purchase.

Chapter

13

ProCAD, Interactive CAD Systems

ProCAD is a fully integrated circuit design package that includes schematic capture and PCB layout in one $695 program. For $100 more, you can buy *ProCAD Xtra,* which includes EMS support with design rule check software, and another $795 nets you an autorouter. Yes, this is a nickel-and-dime package, but it ends at $1,995, and for your money you get more performance than most modular systems provide.

At least 128K of extended or expanded LIM 4.0 EMS memory is needed for *ProCAD Xtra* and *ProCAD Xtra-XL* to load. For large, involved circuits, you will want to spend another $200 (or $250 for upgrading later) and get the *ProCAD Xtra-XL* version, which lets you use up to 8 MB of LIM 4.0 EMS memory. Otherwise, you will have to put up with slower operation while *ProCAD*'s memory manager shuffles data between the hard disk and an overlay cache that it automatically sets up in the shadow RAM area between the PC's basic 640K and 1 MB. Fortunately, *ProCAD* uses memory economically though clever data base memory management that renders a 512K hard disk device file into just 20 bytes of RAM.

While a hardware protection key (dongle) is not needed, the program is copy protected using the old Lotus scheme that transfers the protection code from floppy to hard disk so that it can only be used on one PC at a time. Obviously, a hard disk is required. A math coprocessor is not required, but Interactive CAD Systems strongly recommends its use.

Unlike the other integrated packages in this book, *ProCAD* does not have a common menu, forcing you to call up the utilities from the DOS prompt. However, the program supports both schematic capture and PCB layout from the same program, which means all you have to do is change the data base file and library to switch from one to the other. Unfortunately, you have to quit to program to generate the netlists that link the two together. Learning and using *ProCAD* can be frustrating too because everything is user programmable, meaning it is easy to make serious mistakes. But if you do not like the way *ProCAD* does things, you can create your own menus and command structures.

Schematic Capture

As mentioned, the company offers a few different software versions of *ProCAD*. The one we examined is *ProCAD Xtra*, which is priced at $795.

It comes with nine libraries, only one of which may be active at any time. For an added $150 you can buy seven optional libraries (separately they cost $35 each), which brings the number of components available to about 1,500 devices. With all the optional libraries installed, *ProCAD* requires more than 6 MB of free hard-disk space. The library includes a strong storehouse of surface-mount devices, with specific commands for working with them. DeMorgan equivalents are not provided, but you can create your own using the library editor, as well as any other devices you find lacking.

Although only one component library is active at a time, it is relatively easy to select a component from another library or create a new symbol without ever leaving the worksheet. Operations, mostly with a mouse, are easy, being only a few mouse clicks away for nested commands. Pop-up menus further simplify matters. For example, calling up a library produces a scrolling menu that lists all the devices in that library; highlighting the part and clicking the mouse button moves the part from the library to the schematic. You can also type in a device name at the command line, if you wish.

To draw wires, you click on the place command, then click on wire to get into its submenu. You then position the cursor on a device's pin, click, and draw wire out by moving the cursor. At a point where you wish to change direction, you click the left mouse button (or press <ENTER>) and continue pulling out more wire. When you are finished, you press the rightmost button or hit <ESC>. If you cross wires, a warning beep is sounded and an X is displayed at the "short" location to alert you to a possible short circuit. You then have the option of continuing along or placing a dot on the crossover to make the connection.

ProCAD's first-rate schematic capture editor has all the standard bells and whistles, plus some. For example, there is a Width command that allows you to increase the default width of a wire, circle, text height, and more in whatever increments you wish. There are even two interpreter shells to choose from: a two-letter mnemonic and a VMS DCL type. Rotation of a symbol features extra power; it can be done in 1-degree increments rather than the 90-degree choices offered by most programs. It also has an adjustable

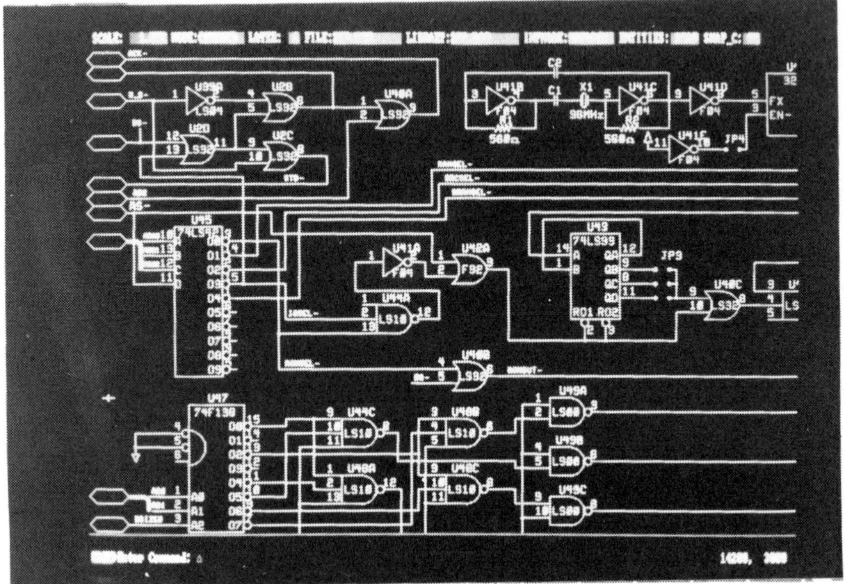

Figure 13-1 A *ProCAD* schematic capture screen.

automatic save feature. And zoom magnification is unlimited. Furthermore, lines can be drawn at any angle.

Editing features include rubberbanding, snap-on grid, dragging, scaling, block moves, undo, undelete in hierarchical steps, macro command, step and repeat, automatic ground and voltage connection, autopan, mirror both X and Y axes, and find. A strong design rule checker checks for opens and shorts, although most of its power is reserved for PCB work.

Component designations (like U1, U2 . . . , R1, R2 . . . , etc.) are automatically assigned in the sequence in which the parts are drawn. If a part is deleted, the program automatically renumbers the references to eliminate the gap. Therefore, there is no possibility of duplicate component numbers.

ProCAD's schematic capture program can generate netlist support for a lot of programs, including *Futurenet, Simulog,* and EDIF, among others. There is also support for *AutoCAD* DXF files and *PSpice.*

PCB Layout

The link between the three functions, schematic capture, PCB layout, and autoroute, is very tight. Forward and back annotation is supported, but the changes to the netlists must be done manually from the DOS prompt using an ASCII file editor. The PCB software also recognizes *FutureNet* netlists.

The program comes with nine library modules, two of which contain device outlines and PCB pads (about 50 devices in all). But the parts count is so skimpy that you will want to consider obtaining one or more of the optional libraries, which sell for $35 each or six for $150. But like the schematic capture software, only one library is active at a time, forcing you to make frequent library changes.

While there is no automatic parts placement routine, device outlines can automatically be called from the schematic netlist one at a time for manual placement. A board outline is required prior to netlist parts placement. Protected zones may also be defined at this time. Maximum board size is 64 64 inches. As the devices appear, you can locate them anywhere you want (even outside the board) except in the protected zones, skipping those you wish to deal with later. Parts may be rotated in one-degree increments both during and after placement. Unfortunately, there are no initial placement aids. A ratsnest is available only after all the devices are on the circuit board.

The optional autorouter ($795 for regular and $995 for EMS) is an advanced gridless router called a probe router that can actually look ahead to see whether or not placing

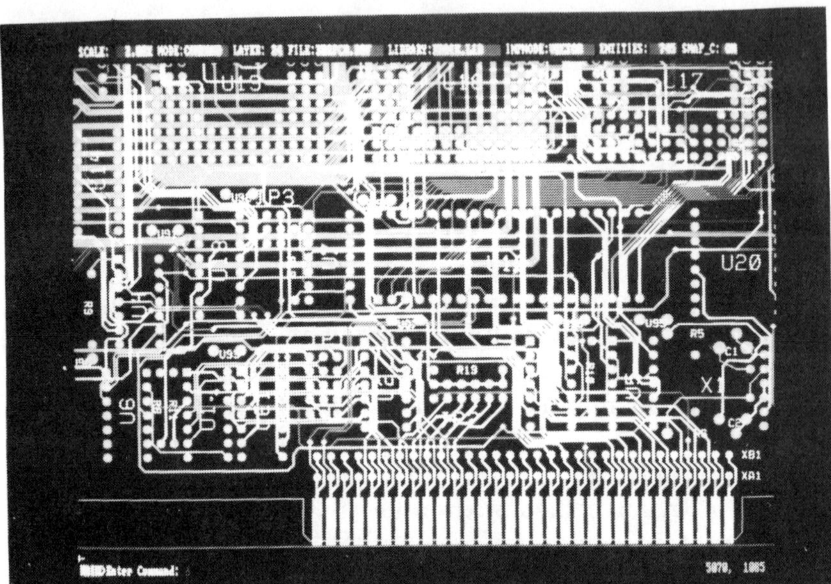

Figure 13-2 A *ProCAD PCB* layout screen.

the current track will block the placement of a subsequent track. Altogether, the router can route two layers at a time using six pass options, each of which can be programmed for a different route parameter and track size, and the router may be stopped at any time for manual intervention. Both track width and grid size are infinitely variable. Routing tests of quite complex circuits using only two layers proved quite impressive, scoring 90 percent on a design with 200 logic ICs (designs of this magnitude generally require four layers). A rip-up router is available for $2,495.

Tracks are laid down manually by clicking once on the mouse and advancing the track to its destination or turning point. You can change layers during track routing, with automatic via placement, but you cannot change track width. *ProCAD* supports nine copper layers out of the box, but you can define another 50 layers as copper for a total of 59. A ground plane may be assigned to any of the layers and connected to any net. Connectivity is maintained when moving already placed tracks.

ProCAD has autopan, but only on three sides because bumping against the bottom nets you a menu rather than a screen scroll, which means your screen is on a continual up-climb until you forcibly bring it back to earth using the manual pan. Autopan is not available during part or track placement, but you can activate manual pan from the keyboard. Zoom is infinitely variable — seemingly down to the atom— and is also available from the keyboard during placement. Anything you can do on the screen can be done as a block. Context-sensitive help is available.

Only *ProCAD Xtra* and *ProCAD Xtra-XL* have design rule checking software, and it only checks for broken tracks and air-gap violations. To save time when checking a modified layout, you can limit your verification to a select portion of the board.

All of the standard netlists and manufacturing files are generated within *ProCAD*, including solder and SMD paste masks and the usual range of silkscreen mats and layout documentation. However, a Gerber netlist generator costs extra ($75), as does N/C Drill ($100) and DXF ($250).

Printer choice and control is admirable, ranging from nine-pin dot-matrix to laser printers and all manner of plotters (including Hewlett-Packard, Houston Instruments, and IBM). Software control lets you choose plot layer, choice of plotter pens or multiple pens, and COM port. Of course, some dot-matrix patterns may only be suitable for proofing unless you watch your line widths and spacing.

Advanced ProCAD Features

ProCAD is not the easiest program to learn and use, owing to its great power. Therefore, the following assists are most welcome. An extensive help menu is a good assist, with a listing by name and syntax coming up in a window. Typing the name of the command you want information on brings up a complete explanation of the command. Help menus can be defined by the user, too. There are also 40 function-key control set-ups. They, too, can be user-defined. Macros and toggle menus also add to the program's productivity.

However, as one becomes more familiar with all its working through heavy use, you may want to try the "Quickset" feature that turns off some of the visual aids and prompts to speed up its use for expert users. Quickset has three settings: novice, intermediate, and advanced.

Advanced users will find that the use of layer assignment colors (*ProCAD* uses 16 colors) are excellent visual aids. For instance, schematic symbols are green, text is red, wire is yellow, highlighting is white, and so on. A status line at the top of the screen keeps track of command

```
              :::: PROCAD Parameter Configuration Ver 3.1 ::::
              (c) 1986-1990 Interactive CAD Systems.  05/30/90
A  --  ICS Product Configured                              PROCAD
B  --  PROTERM Communication Port                          n/a
C  --  PROTERM Initial Baudrate                            n/a
D  --  Default Screen type                                 Normal
E  --  Pointing Device Support                             Mouse
F  --  Mouse type Support                                  MSystem/Logitec
G  --  Mouse Sensitivity threshold                         1
H  --  Mouse/Tablet Communication Port                     2
I  --  Monochrome or Color Support                         Color
J  --  Color lookup table support                          Yes
K  --  Errors/Messages Highlite Support                    Yes
L  --  Grid type Support                                   Dots
M  --  Display of grid tracking marker                     No
N  --  Working file drive prefix                           Default_Drive
O  --  Aspec ratio support for circles                     Yes
T  --  Database Directory Path                             .\
U  --  Library Directory Path                              .\
V  --  Help level                                          Novice
P  --  Select Display & Printer Drivers
S  --  Save Changes & exit
Q  --  Abandom changes & exit
Enter Command: V
Default Help Level [N]ovice, [I]ntermediate, [E]xpert (N) ? _
```

Figure 13-3 Quickset lets expert users reduce the number of visual aids and prompts shown on the screen.

Table 13-1 Circuit Design Rating Summary

Schematic Capture:

Drawing	Editing	Library	Netlist Support	Ease of Use
excellent	excellent	good	excellent	fair

PCB Layout:

Place	Route	Library	Netlist Support	Ease of Use
fair	excellent	fair	poor	fair

settings, such as layer number. There are 99 layers available (think of them as Mylar overlays). A color-coded layer bar at the bottom of the worksheet allows you to click on a color to execute its representative command, like activate a library module, and so on.

Mouse click steps can be reduced by taking advantage of this color code. For example, you can repeat a command and subcommand by simply clicking the left mouse button or pressing <ENTER>. In other instances, the program automatically switches the highlight bar to the proper subcommand when the next operation is obvious.

Among *ProCAD*'s many utility features is a log function that records everything you do so that you can backtrack in the event you want to search out an error you made. Deletions are saved, too, which can be recalled at any depth if required.

ProCAD in any of its many versions is an excellent circuit design package. While it may not be the most automated and certainly not the easiest to use, it has all the features you will ever need for even the most complex design, and it is well suited for the most demanding circuit design task — provided you buy all the options.

Chapter

14

Protel, Protel Technology

Protel is a dominent player in the circuit design industry and serves up a trio of schematic capture and PCB layout programs for PC use. They are *Protel-Schematic* schematic capture, *Protel-Autotrax* PCB layout, and *Protel-Easytrax* PCB layout. All are easy to learn and extremely easy to use. Prices range from $450 to $1,295.

Protel-Schematic

Protel-Schematic is an extremely easy-to-use schematic capture program and among the most versatile. It is priced at a reasonable $495 and comes with a hefty component library.

Protel-Schematic will make use of up to 4 MB of LIM memory if available, but it is not needed for most layouts because of the program's relaxed hardware requirements, which are 640K of RAM and two floppy drives. A math coprocessor is of little value because the program uses CPU-intensive integer processing rather than floating-point math coprocessor calculations. The included hardware protection key (dongle) must be installed for the program to load.

The component library contains over 3,000 parts distributed among 14 library modules, 10 of which may be active. There is a good balance between analog and digital parts, and components can be called from the library by either entering its generic name from the keyboard or by clicking on the part from a scrolling library menu. Oddly enough for a library of this size, DeMorgan equivalents are not supported.

Device reference and multiple-package designations are assigned at the time the part is placed on the worksheet, and there is a warning message that prevents you from assigning the same reference to more than one device. The program remembers the last device placed on the worksheet, permitting repeated component placement without having to reenter the part name, but you have to enter a reference name before the software places it into position. Devices have only two reference lines, neither of which can be moved. However, the size of the font is variable between 8 and 20 points, and the schematic reference name can be hidden, leaving just the part value.

Parts can be rotated and mirrored either during or after placement. There is also a pin editor that lets you add, delete, or change the position of the pins associated with a component. The pin editor can be used as a quick way to make a one-of-a-kind part without having to use the library editor, but parts made this way have to be redrawn for each new schematic because they are not transferable between files.

Wire placement is simply a matter of clicking the mouse button once to start the line and clicking it a second time to place the line and change its direction. *Protel-Schematic* also lets you draw 45-degree diagonals in the ortho mode.

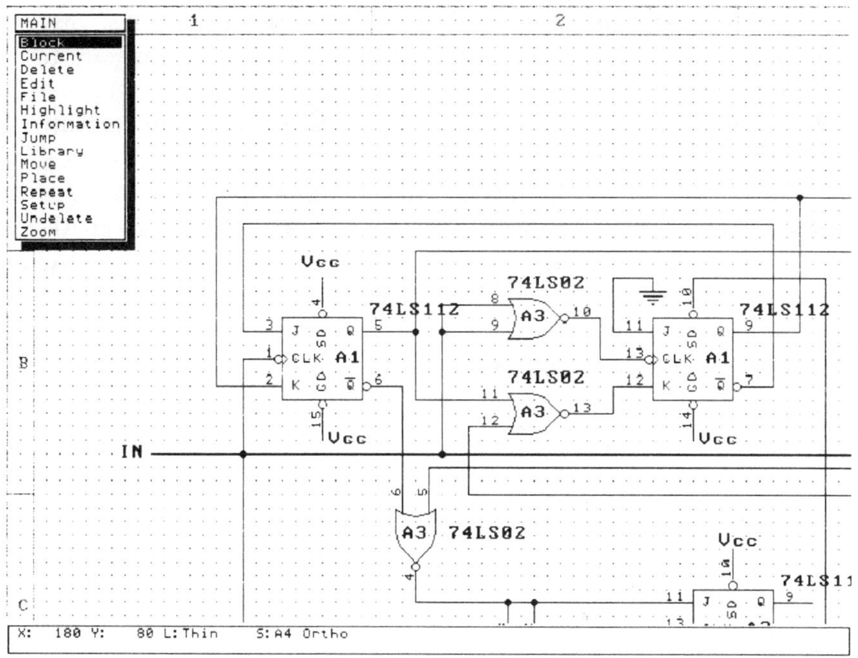

Figure 14-1 A *Protel-Schematic* schematic capture screen.

Protel-Schematic has a macro recorder that records and saves your every keystroke or mouse click when activated. There is no limit to the size of the macro or how many macros can be created. Macros can even be nested in batch file fashion so that starting one macro runs a sequence of queued macros.

Autopan is supported, but it jumps rather than scrolls. The grid and grid snap cannot be turned off, but you have full control over the screen display colors, which means you can set the grid to the same color as the background and effectively hide it if it bothers you.

The editing features are excellent, with single object move, delete, and jump, plus block editing. There is no undo, but there is an undelete command that lets you undelete as many items as you wish in the reverse order they were deleted.

Only the library symbol editor has a help screen, and that is limited to a succinct summary of Chapter 24 in the user's manual. All netlists are in proprietary *Protel-Schematic* format.

Printer and plotter hardcopies are both of excellent quality. Hardcopy output is done using an external utility that gives you complete control over all aspects of the printer or plotter, including drawing orientation, fit drawing to page, and page offset, to name but a few.

For pure hardcopy drafting quality and editing freedom, *Protel-Schematic* is rivaled only by high-priced CAD workstations. However, its lack of support for formats other than Protel limits its PCB layout application to Protel products only.

Protel-Autotrax

Protel-Autotrax is a moderately priced $1,295 PCB layout program that has both automatic parts placement and autorouting. Easy to use, the program recognizes several schematic capture netlist formats and is ideal for fast-to-copper prototype and limited production boards. But it lacks the routing conveniences found in more sophisticated PCB programs.

Protel-Autotrax will make use of up to 4 MB of LIM memory if available, but it is not needed for most layouts because of *Protel-Autotrax*'s relaxed hardware requirements, which are 640K of RAM and two floppy drives. A math coprocessor is of little value because the program uses CPU-intensive integer processing rather than floating-point math coprocessor calculations. The included hardware protection key (dongle) must be installed for the program to load.

Because the program has both automatic parts placement and autorouting, the program is extremely easy to learn and use. A few mouse clicks and you are on your way, unless of course, you want or need to do some of the board layout on your own.

Protel has designed this package to be fully compatible with its other engineering programs, and the mesh is flawless. Although there is no back annotation to the schematic, there is a NETCHECK utility that compares schematic capture and PCB netlists and reports all discrepancies. *Protel-Autotrax* provides a utility for converting *OrCAD, Schema,* and *PADS-PCB* netlists into Protel format, all of which support automatic parts placement and therefore device outlines, but be prepared for heavy-duty manual editing of the netlist files because the device outlines lose more than a little in the translation.

The component library contains outlines for about 170 devices. The inventory is evenly divided among passive, DIP, and connector footprints, with a goodly number claimed by SMD devices. The outlines are not attached to any particular part number or device and are used over and over again in a layout without regard to the actual device.

Before placing components, the board outline must be defined. Protected zones that prevent the placement of parts or tracks within the zone may also be defined at this time. Maximum board size is 32 × 32 inches.

The automatic placement routine is on par with *EE Designer's*, which is sufficient but not something to write home about. Parts that are manually positioned prior to or after automatic placement are automatically glued in place, allowing you to run the placement routine again for better device positioning. Parts may be rotated or moved to the

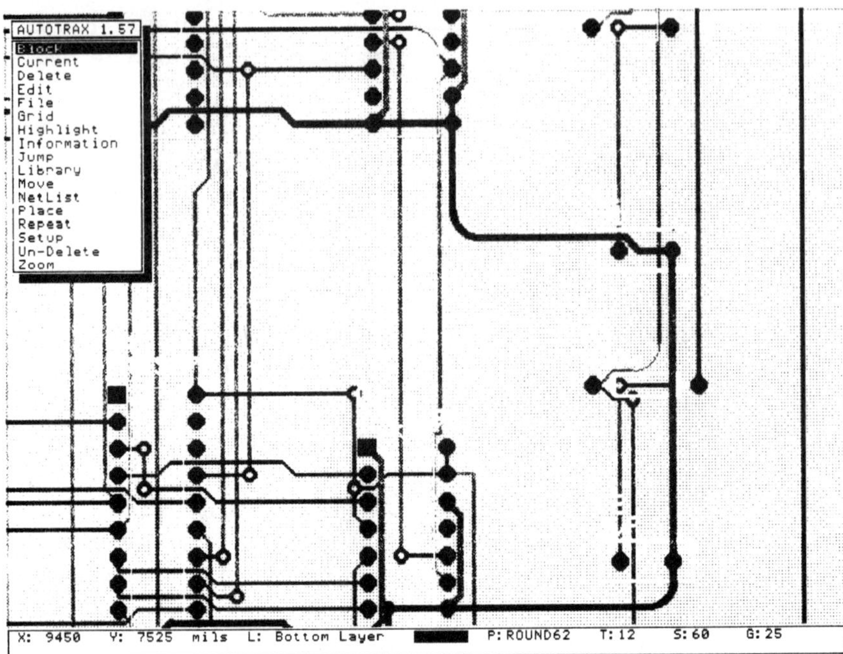

Figure 14-2 A *Protel-Autotrax PCB* layout screen.

flip side for SMD layouts after initial placement. A ratsnest is available for placement.

The Lee-based autorouter is costed, which allows you to dictate the types of routing routines permitted. The router has 10 routing variables, all of which may be activated in any combination. In addition there are six optimization functions, like via minimization and loop removal, which may also be activated. The costing routines are implemented in order from the simplest to the most exhaustive, with the optimization routines running last. There is also a smoother routine that optimizes manually placed traces without violating the designer's routing intentions.

Although all six copper layers can be included in the autoroute, only one track width can be routed at a time. To lay down tracks of different widths, you must specify the net to be routed, enter the track width, then route the net to completion and proceed to the next net and do the same. For example, to have thicker power supply and ground tracks you have to do three routings: one for VCC, one for GND, and one for the remaining signal nets. The router grid is variable from 5 to 1,000 mils, and the track width is variable between 1 and 255 mils. *Protel-Autotrax* interfaces with Protel's $1,495 Traxstar rip-up router.

Manually routed tracks are laid down by clicking the mouse button once and moving the track to its destination and clicking again to change direction. Tracks may be ortho, free style, or curved. However, you cannot change track type on the fly. Connectivity is maintained when a track is moved, with the track stretching or contracting to accommodate its new location. There is an undelete command that systematically replaces deleted tracks in the order they were erased back to the beginning. Only two ground planes are supported, but they may be attached to any net.

Screen pan is automatic, but not during automatic placement or autorouting. The 100:1 zoom, on the other hand, is available during placement and routing, but is not available when a pull-down command menu is displayed. Block functions include move, rotate, copy, inside and outside delete, hide, and save. Macros are supported and may be nested.

The design rule check command reports spacing violations, broken or unrouted nets, missing components, and extra pins. The coordinates and type of error are scrolled on the screen and saved to file.

Protel-Autotrax produces both Gerber and N/C drill files. Other output files include silkscreen, solder, and SMD paste masks. The component placement outlines and nomenclature are combined in one mask.

Final artwork may be produced on either a dot-matrix or laser printer, and with care can be used as final artwork. PostScript format is supported, and the artwork can be scaled to any size. Several plotters are also supported with plotter adjustments set from the software.

Protel-Autotrax is a medium-priced PCB layout program that tightly integrates with the other members of the Protel circuit design family. Its strength is the size of its component library, the many costed functions the autorouter provides, and the fact that a rip-up router is available. It is ideal for small production runs and prototypes, but lacks the power needed for serious multilayer work.

Protel-Easytrax

Protel-Easytrax is Protel's low-end PCB layout package. The program is essentially the same as *Protel-Autotrax* — except that it is totally manually operated with no contact with the outside world save its output files. However — and this is a big positive — it does have a very sophisticated point-to-point autorouter that may well justify its $450 price tag for occasional users. I have personally used it to create a one-of-a-kind, highly specialized PCB layout simply because of its simplicity, rather than hassle the procedure of drawing a schematic diagram for input to the PCB layout software.

The program is easy to learn and use, but everything on the board is put there by hand, one part or track at a time, because the software doesn't interface with any schematic capture or router program — or any type of netlist, for that matter. It is strictly a stand-alone package.

The component library contains about 100 device outlines, which is 50 percent less than *Protel-Autotrax* but

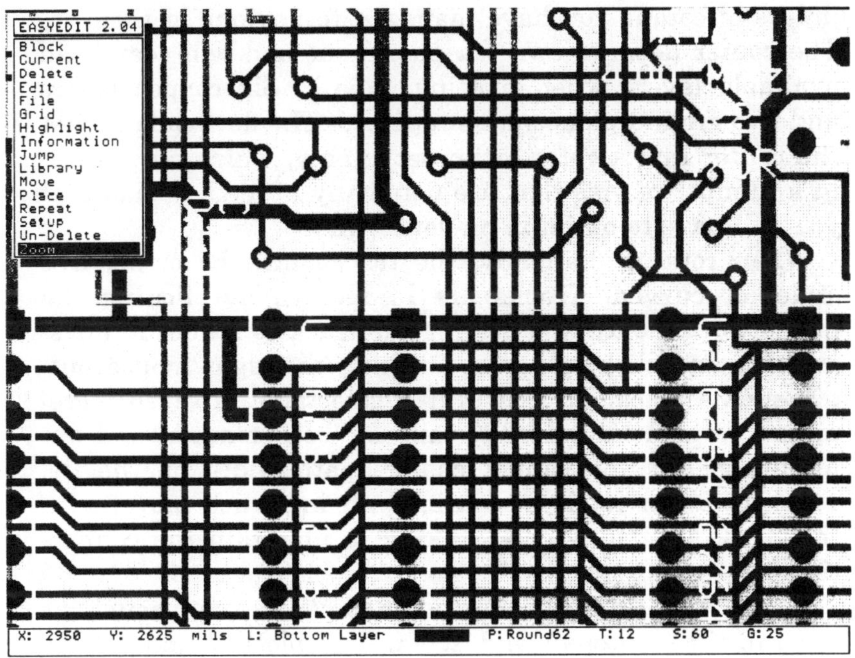

Figure 14-3 A *Protel-Easytrax PCB* layout screen.

decent. Missing are SMD outlines and several connectors. Like *Protel-Autotrax,* you can add your own parts or create new libraries using the library editor.

Parts are placed on the board by either typing their name from the keyboard or selecting the device from a scrolling library menu. While you can rotate the part during and after placement, parts may only be placed on the top side of the board, preventing the use of SMDs. Once a device outline is selected, it may be repeated as many times as you wish. Maximum board size is 32 × 32 inches.

Tracks may be placed manually or by using the pad-to-pad autorouter. The single-minded heuristic router is very clever in finding a way between the two specified pads, unless of course you have managed to box one of them in. The router does its own via placement and will search all available layers when plotting a path. Six copper layers and two ground planes are supported. The downside is that the router investigates ortho paths only, which may result in a number of vias that are not really needed — vias that you will have to optimize on your own.

When routing manually, the tracks may be ortho, free style, or curved. Tracks are laid down by clicking the mouse button once and moving the track to its desired destination and clicking again. Clicking the right mouse button ends a track. Track width is selectable in seven increments between 10 and 100 mils for both manual and autorouting, but you cannot change path width on the fly. Track connectivity is maintained during track move, with the track stretching or shrinking to maintain the connection.

Like *Protel-Autotrax,* nested macros are supported, as are full-time autopan and zoom. The block functions are also identical in every respect, and the undelete command still replaces tracks that were erased back to the beginning of the current drawing load. However, there is no design rule check.

Protel-Easytrax's only contact with the outside world is via its output files. *Protel-Easytrax* files can be converted into *Protel-Autotrax* files, giving the *Protel-Easytrax* designer the advantages of *Traxstar* and other Protel goodies, but *Protel-Autotrax* files cannot be changed to *Protel-Easytrax* files because of *Protel-Autotrax*'s superior editing capabilities.

Like *Protel-Autotrax, Protel-Easytrax* produces Gerber and N/C drill files, as well as silkscreen and solder masks. *Protel-Easytrax*'s printer and plotter interface are identical to *Protel-Autotrax*'s, which means you can produce final artwork at scaled sizes on dot-matrix or laser printers and several pen plotters.

Table 14-1 Circuit Design Rating Summary

Schematic Capture: Protel-Schematic

Drawing	Editing	Library	Netlist Support	Ease of Use
excellent	excellent	excellent	poor	excellent

PCB Layout: Protel-Autotrax

Place	Route	Library	Netlist Support	Ease of Use
excellent	excellent	good	fair	excellent

PCB Layout: Protel-Easytrax

Place	Route	Library	Netlist Support	Ease of Use
poor	fair	good	none	excellent

Protel-Easytrax is a complete PCB layout program that is easy to learn and use. But because it lacks support for an input netlist, you will be doing all the work by hand, even when calling on the autorouter to make pad-by-pad connections. Still, its fantastic support of printers, Gerber and N/C files plus production masks at no extra cost make this program a good choice for an advanced hobbyist or the start-up business on a budget.

Chapter

15

Schema, Omation

Omation sells three circuit design packages: two schematic capture programs and one PCB layout program. *Schema III* is their mainstream schematic capture program, which sells for $495. A student (university) version of *Schema III* called *Schema-Quik* sells for $99 individually or $399 for ten (making them only $40 each).

Schema-PCB Layout is a fantastic PCB layout program that they OEM (buy) from PADS Software, Inc., of Littleton, MA. It contains everything, including a kitchen sink, but sells for a hefty $2,395 or more by the time you buy all the modules needed to make it do its marvelous magic.

Schema III

Schema III is a $495 schematic capture program that is easy to learn and use, but it lacks some of the drawing and editing skills other programs in this price range have.

Schema III's component library consists of about 1,000 parts divided among nine library modules, with heavy emphasis on TTL devices. Many of the objects in the library are outlines and headers that let you easily piece together custom components using the library editor. All nine library modules may be active at the same time.

Components can only be called from the library by entering their name from the keyboard, and you may assign complete schematic references at this time, including part value and up to six lines of additional text. However, *Schema III* cannot detect if the same device reference is used for more than one part, making it your responsibility to avoid such errors. Reference lines cannot be moved or hidden.

There is no component repeat function. Each device placed on the worksheet must be called up individually. However, *Schema III* has a copy command that lets you make multiple copies of a part already on the drawing, but you will have to use the image editor to change the part reference names because they are also duplicated.

Schema III has no parts orientation editor. Instead, the different orientations are stored as separate devices in the component library, as are the DeMorgan equivalents. Each orientation has its own name, which you specify at the time you call up the part. However, after a part has been placed on the drawing, you can use the rotate command to step it through its various library orientations.

Wires are drawn by holding down the right mouse button and scooting the rodent to the desired location. Releasing the button places the wire on the worksheet; pressing the button again starts a new wire. Diagonal lines drawn at any angle are made by selecting the diagonal line option from the menu. There is no bus drawing or editing function, but buses can be created using the thick wire and arc menu options in conjunction with a repeat command that places the breakouts at user-defined intervals.

Autopan is smooth scrolling and the speed of the scroll is adjustable, as is the grid's dot spacing. Unlike most

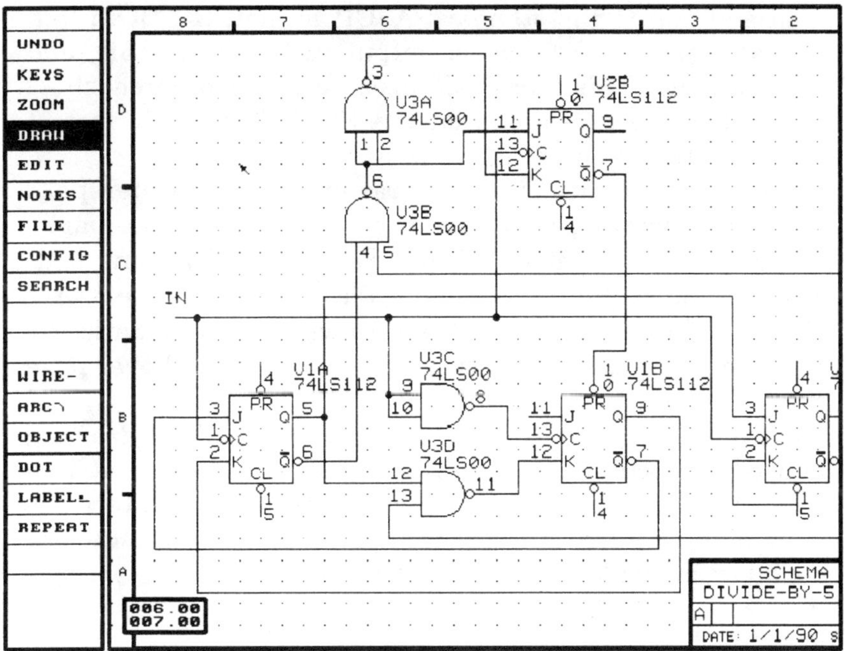

Figure 15-1 A *Schema III* schematic capture screen.

schematic capture programs, you can choose your zoom size from a menu rather than having to step through each zoom level one by one.

The editing features are average and consist of move, delete, and copy. Objects may be edited individually or as blocks.

Schema III has an on-line help screen. Help is context sensitive (matching the help message to the current command) and is available for both the screen and library editor.

Schema III supports several popular netlist formats. Among the 14 netlist output formats supported are SPICE and *EE Designer.*

In addition to the standard ASCII format, the BOM (bill of materials) netlist can be output in Lotus, dBASE, and SDF formats. This allows the user to perform spreadsheet and data base operations on the BOM data, such as cost analysis and inventory control.

While you cannot size the drawing to the printer page (an A-size drawing only uses about one-third of the page), you can take advantage of the printer page by using a C-size worksheet for 8½-inch printers and an E-size worksheet for 15-inch printers when drawing the schematic. By contrast, the plotting utility gives you full control over page sizing, reduction, and offset.

Although *Schema III* does not have the refined drawing and editing skills demonstrated by some packages, it is a well-rounded schematic capture program that interfaces with many popular PCB layout programs and is quite acceptable for light production work.

Schema-Quik

Schema-Quik is Omation's student-version schematic capture program of *Schema III*. But low price does not mean low performance: *Schema-Quik* does everything *Schema III* does, but on a smaller scale. Like *Schema III*, the drawing and editing features are more limited than some mainstream schematic capture programs, but at $99 (as opposed to *Schema III*'s $495) they are a lot easier to live with.

Unlike *Schema III*, *Schema-Quik* does not support LIM memory, a fact that limits *Schema-Quik* drawings to about 200 ICs or less. A hardware protection device is not required, and the software is not copy protected.

The program is easy to learn and use. Commands are called from simple menus that are only one level deep; no winding paths here. A context-sensitive help screen that matches the help message to the highlighted command is available.

Autopan is smooth scrolling and the speed of the scroll is user adjustable, as is the spacing of the grid dots. Zoom has an 8:1 range that can be invoked from a menu or the function keys.

Schema-Quik has a single component library that comes with about 1000 devices, mostly TTL chips with similar part numbers like 74LS00 and 54LS00. A count of different devices is something closer to 100. However, you can add up to 3400 additional devices of your own making using the library editor, and you do not have to quit the schematic capture program to use the editor.

Components are called from the library by typing in the devices name from the keyboard. The drawing editor has a browse feature that lets you look through the library and select a part for placement by clicking on it.

You can assign complete schematic references at this time, including part value and up to six lines of additional text. Reference text can be modified at any time using the image editor, but the reference lines cannot be moved or hidden.

You can use the draw-repeat or the copy command to place more than one part of the same kind. When using copy, however, you will have to use the image editor to change part references because they too are duplicated.

Wires are drawn by holding down the left mouse button and scooting the rodent to the desired location. Releasing the button places the wire on the drawing; pressing the button again starts a new wire. Diagonals are not supported, and neither are buses. There is a thick wire drawing function that lets you create the look of a bus, but it is not electrically correct and appears as one wire in the output netlists.

Editing features are average and consist of move, delete and copy, and objects can be edited individually or as blocks. Absent is mirror imaging.

Schema-Quik supports several popular netlist formats for use with PCB layout and circuit simulation software.

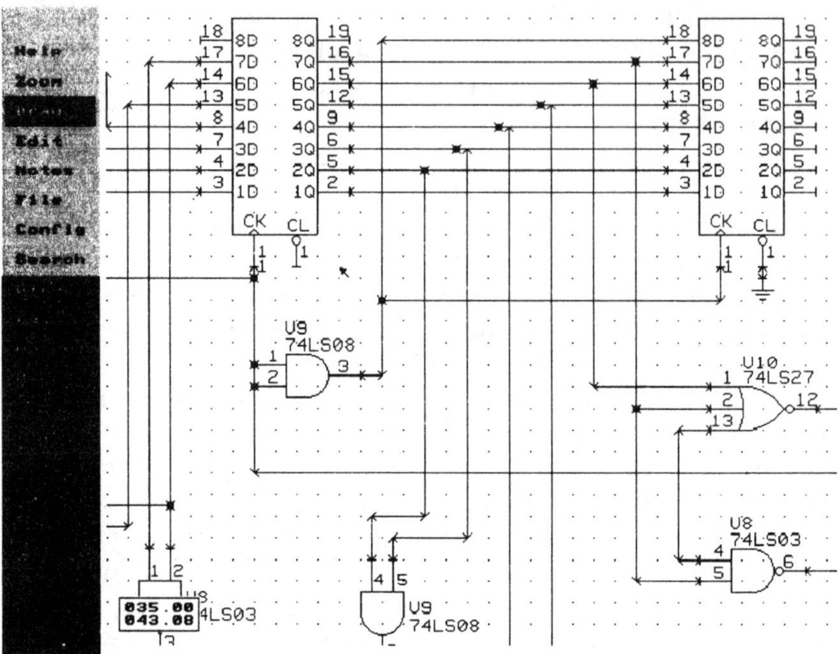

Figure 15-2 A *Schema-Quik* schematic capture screen.

Among the 14 netlist formats supported are SPICE, *EE Designer,* and *PADS-PCB.* In addition to the ASCII form of the Bill of Materials (BOM), there are Lotus, dBASE, and SDF formats that can be imported into a spreadsheet or data base for cost analysis and inventory control.

There are two DOS-based design rule checkers that look for and report on floating inputs, duplicate part references, shorted outputs, and dangling wires.

While *Schema-Quik* supports a large number of printer types, the program has little control over the final hardcopy. That means you will end up doing a lot of taping if the drawing is larger than your printer paper. By contrast, the plotter utility gives you full control over page sizing, reduction, and offset. Both utilities are run from DOS.

Although *Schema-Quik* does not have as many refined drawing and editing capabilities as other programs in this review, it is a well-rounded schematic capture package that interfaces with many popular PCB layout and circuit simulation programs. Its only real drawback is the small component library, but for $99 you can afford to spend some time tailoring the library to your needs.

Schema-PCB Layout

For real PCB layout power, it is hard to do better than *Schema-PCB Layout*. It supports automatic placement, three autorouters, and a trove of editing tools. But brace yourself, because although the program lists for a low $975, you will end up spending at least $2,395 by the time you piece together the packages needed.

The program will support up to 8 MB of expanded RAM. A hard disk is required, and the supplied hardware protection key (dongle) must be installed before the software will load.

Both the schematic capture and PCB layout programs are accessible from a common main menu. The PCB software is easy enough to learn, but tedious to use because the commands are stacked in a hierarchic structure that has you accessing layer after layer when searching for a command and when returning to the main menu. Although you can use the mouse to click on a command, it is actually faster to use the keyboard.

Schema-PCB Layout is tightly linked with *Schema III* and *Schema-Quik* schematic capture programs, with both forward and backward annotation supported. *FutureNet* netlists can also be read, but not without some manual editing of the netlist. Two optional programs for converting P-CAD or Recad-Redac netlists into *Schema-PCB* format are available for $995 and $1,500, respectively.

The library contains outlines for 159 devices, including several SMDs. The library also electrically links these outlines with 2160 components, mostly ICs, that the program uses for automatic placement and package editing. However, the library is a bit touchy about how you specify your parts in the schematic capture. Although *Schema III*'s library contains the 7446 chip, *Schema-PCB* would not accept it unless you request a 54L46.

Schema-PCB Layout has an excellent automatic placement program that even includes a routine for properly placing decoupling capacitors. But get ready to up the ante again, because the utility costs an extra $350. Before you can extract device outlines from the schematic netlist, you have to define both a board outline (up to 32 × 32 inches) and a parts matrix field. Practically, the matrix should contain the same number of fields as the number of parts to be placed; if it is less, more than one part is placed in a matrix block. The matrix is not sized to the board so that you can place sections of the board according to device type, such as memory chips and the like. While this is a great asset, it takes some planning, because making a mistake on the matrix sizing may net you a board with all the components huddled in a corner or scattered to the four corners of the globe. A matrix is not required for manual placement.

Parts may be rotated or flipped to the solder side of the board for SMD layout after auto placement. Any part can be glued in place and the automatic placement run again for improved placement. A ratsnest is available.

The router options are nothing short of fantastic, but again get ready to pony up more money. The standard autorouter ($750) uses a mix of heuristic and Lee algorithms that are under full user control. Altogether there are 15 routing options, 10 of which are Lee based, which may be selected in any combination. The autorouter runs through the list beginning with power supply tracks and ending in via optimization. However, you can only autoroute two layers and one track width at a time. Autorouting may be confined to a net or specific area of the board and may be stopped at any time for manual intervention.

The router grid is infinitely variable between 1 and 800 mils and can be changed on the fly. An optional rip-up router that can route 12 layers simultaneously sells for a hefty $3,500 and a shove-aside router tips the scale at $1,495.

For manual routing, the tracks are started by clicking on a pad and moving the pointer to the connecting pad. A ratsnest, which can hide power supply nets, is always visible. Changing the track width for necking down can be done on the fly. Curved tracks are not supported. Connectivity is maintained when moving a placed track. The program supports 30 copper layers, and each can have both a power and a ground plane.

Schema-PCB Layout is one of only a handful of programs that automatically reassigns pin numbers when changing the package type, including changes from DIP to SMD packages. However, the feature is limited to the 2160 devices contained in the component library. Which means for package changes to generic devices, you're on your own. *Schema-PCB Layout* supports gate and pin swapping. Blocks of components and tracks may be moved, rotated, deleted, mirrored, copied, and saved to file. Connectivity to parts outside the block can be maintained using the rubberband feature. Pan and zoom are manual, not automatic. However, both pan and zoom are available for parts placement and routing. Zoom range is 64 to 1.

Schema-PCB Layout has an extremely sophisticated design rule check program. In addition to the usual air gap violation netlist, there is a density map that displays areas

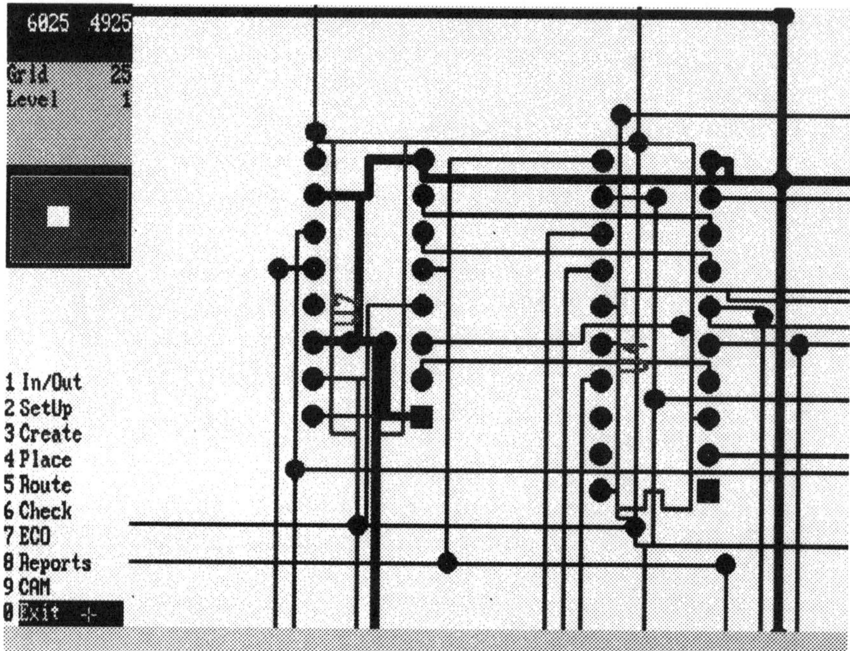

Figure 15-3 A *Schema-PCB Layout* PCB layout screen.

on the board where the concentration of tracks may cause problems and a histogram that shows you how to move devices for better routing.

The program can produce a full range of output netlists and files — at extra cost, of course. The Gerber and N/C drill utilities are $300 each, and the DXF file generator is $495. The only thing you do not have to pay extra for are the layer masks, which includes all 30 copper layers, solder and SMD paste masks, and top and bottom layer silkscreen artwork.

A dot-matrix or laser printer can be used to produce all artwork. Art may be sized to fit the page, scaled to a model size, or printed in actual size. However, a wide carriage

Table 15-1 Circuit Design Rating Summary

Schematic Capture: Schema III

Drawing	Editing	Library	Netlist Support	Ease of Use
good	good	good	excellent	good

Schematic Capture: Schema-Quik

Drawing	Editing	Library	Netlist Support	Ease of Use
good	good	fair	excellent	good

PCB Layout: Schema-PCB Layout

Place	Route	Library	Netlist Support	Ease of Use
excellent	excellent	good	good	fair

printer is required to get it all on one page even at the 1:1 scale. When printed in sizes of 2X or greater, the artwork is good enough for limited production and prototypic purposes. Both Hewlett-Packard and Houston Instruments pen plotters are supported.

With support for 30 copper layers and a bevy of autorouters to choose from, *Schema-PCB Layout* is one of the best PCB board layout programs around. What it lacks in ease of use, it makes up for in features. But you had better get your pocketbook ready, because to get everything you want may cost a tidy bundle.

Chapter

16

Tango, ACCEL Technologies

ACCEL Technologies sells a complete line of PC-based programs called Tango that are used for schematic capture and printed circuit board layout. In this chapter we look at *Tango-Schematic, Tango-PCB Plus*, and *Tango-Route Plus*.

Although each program is individual and must be run separately from DOS, the Tango netlists are complete and permit both forward and backward annotation between the programs. Unlike OrCAD and Schema, there is no common menu that lets you access the three programs or Tango utilities from a single screen.

All three programs require a hard disk and must have the included hardware protection key (dongle) installed in the PC's parallel port before the program will load. Each program has its own key, and the keys may be stacked so that you do not have to constantly remove and replace dongles as you change programs. Expanded LIM EMS memory is required.

Tango-Schematic Series II

Tango-Schematic is an extremely versatile $495 schematic capture program that is easy to learn and fairly easy to use. It comes with a large component library and has excellent printer hardcopy control. But you will need 5 MB of free disk space plus 1 MB of LIM-compatible RAM to use it.

The component library consists of more than 11,000 devices spread among 23 library modules, and includes military as well as commercial TTL devices in both DIP and surface-mount packages. However, several thousand parts are duplicated among the modules, each with the same body outline but a different library name (e.g., SN74LS112AN, SN74LS112AD, SN54LS112AJ, SN54LS112AW). A count of *different* device types reveals a more realistic library size of about 3500 devices.

Components are called up from the library via a Windows-like dialog box. You may either enter the device name at the prompt or choose the part from a menu list. The menu list is the better choice because most library modules are manufacturer specific, with eight modules alone dedicated to Texas Instrument TTL chips, and a generic name like 74LS02 will not get you anything but a "part not found" warning. The manufacturer's full device name is mandatory. Multiple part placement is supported, as is rotate and flip.

Before you can place a part on the worksheet, its library module must be loaded into the worksheet; up to 10 library modules are permitted per drawing. To reduce the amount of disk space consumed by the library modules, which totals 3.6 MB for all 23 modules, you can extract the parts used by the drawing from their related modules and save them in a drawing-specific library module — which now becomes the only library module needed for that drawing.

Part references may be designated either during or after the part is placed on the worksheet. Automatic part naming is also supported, with annotation starting at the upper left-hand section of the schematic and sweeping to the lower right corner in a raster pattern. Parts that have previously assigned references are not changed, and their reference is not duplicated. However, references manually assigned during part placement or editing are not checked for duplication. Each device has two reference lines which may be edited for text and position, but not font size.

Wires are drawn by clicking the mouse button once to start the line and clicking it a second time to place the line and change its direction. You can also draw 45-degree diagonals in the ortho mode.

Tango-Schematic has no pan function. Instead, the infinitely variable zoom is used to place the work area of interest on center stage. Unfortunately, this does not help with the placement of long wire runs because the small zoom magnification needed to view an entire sheet makes wire alignment difficult. Tango has three grids: an absolute grid that is immovable, an invisible grid-snap reference grid that can be moved with relationship to the first, and a visible grid that cannot be turned off but can be toggled between a dot and a hatch pattern. Dot spacing for each grid is individually programmable.

Editing features are excellent, with move, delete, and find among the many editing options offered for single objects, blocks, and buses. There is also a cleanup routine that removes duplicate wires, but does not catch duplicate parts.

Context-sensitive on-line help is available. Unfortunately, *Tango-Schematic* only supports its own netlist format, which ties you to Tango PCB products unless the PCB layout software includes a Tango netlist converter. Netlists can be checked for drawing errors using a design rule checker utility.

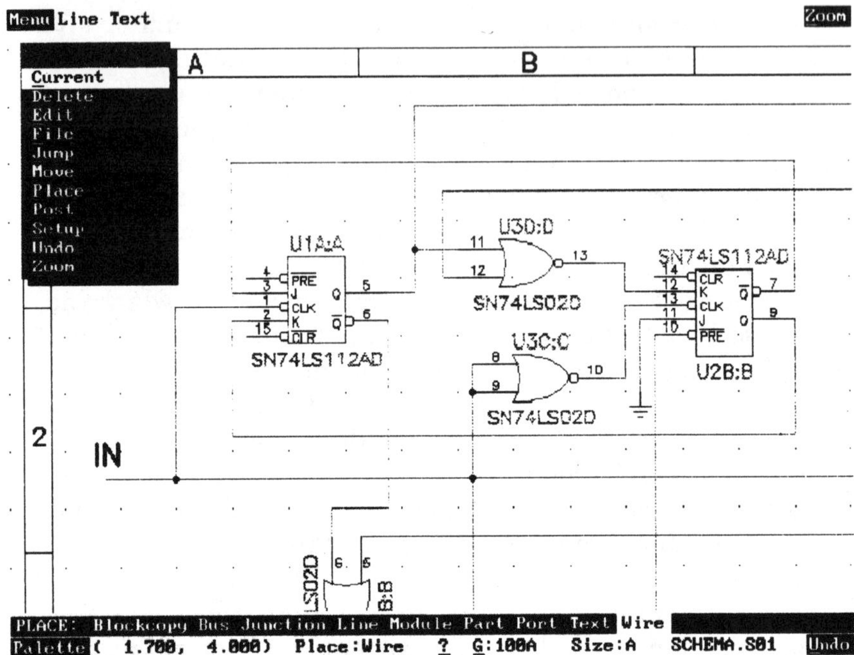

Figure 16-1 A *Tango-Schematic* schematic capture screen.

Printer output is just short of fantastic, with control over all aspects of the hardcopy page, including image resolution, colors for printers that support them, and sizing a drawing to fit the printer page. But this printer driver is no miracle worker, refusing to print an E-size drawing on an 8½ × 11-inch page for obvious reasons. Plotter control is equally good.

Tango-Schematic is an excellent schematic capture program with many drawing features and much versatility. But plan on using it with Tango's PCB layout software if you need a complete PCB design package because the netlist format is not popularly supported.

Tango-PCB Plus

Tango-PCB Plus is an $895 PCB layout program that can be used alone or in conjunction with *Tango-Schematic* and other Accel circuit design products. The program only recognizes Tango netlists, and it cannot extract the device outlines from the schematic. However, the component library size and routing features are above average, and an optional autorouter is available.

The program supports up to 8 MB of LIM 3.2 or 32 MB of LIM 4.0 expanded memory. A hard disk is required, and the included hardware protection key (dongle) must be installed in the parallel port. A math coprocessor is of no value because the program uses integer processing.

The Windows-like interface is easy to learn and use. You can either choose the commands from pull-down menus or via the keyboard. A speed palette offers one-click access to the most common commands and is useful for doing repetitive jobs.

Only Tango netlists are recognized, and then only the pin connections of the list, not the device outlines. Consequently, there is no forward or backward annotation of the schematic and PCB layout like the more expensive *Tango-PCB* 2.0 program offers.

The component library consists of three modules: a main library holding 107 general-purpose devices, a 120-device connector library, and an SMT library with 90 SMD outlines. When browsing the contents of a library, the actual footprint is shown on the screen, eliminating the need for a pattern reference manual.

Since components cannot be extracted from the netlist, you are required to call them from the libraries and manually place them on the board. A scrolling library list is available for placement. Maximum board size is 32 × 32 inches.

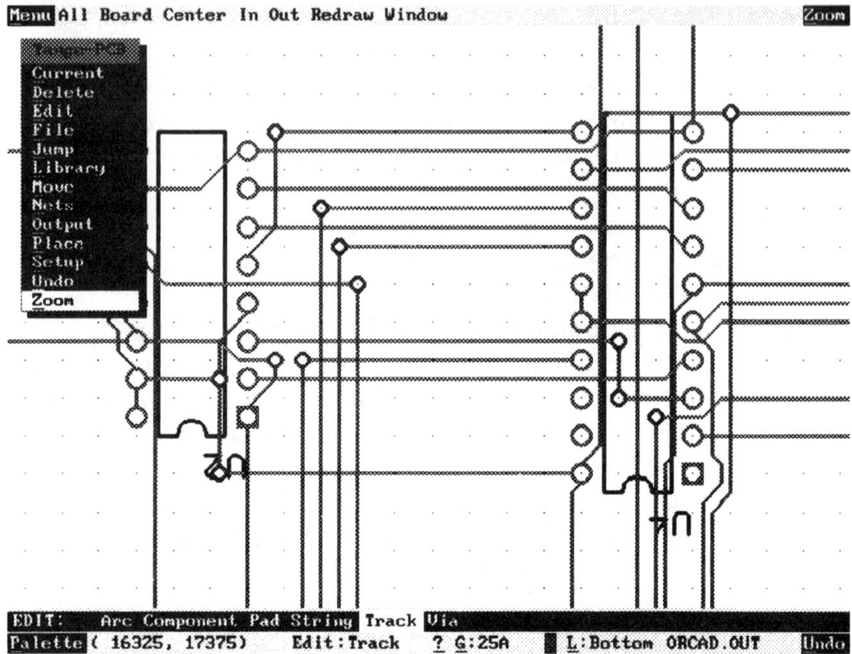

Figure 16-2 A *Tango-PCB Plus PCB* layout screen.

During placement, you can rotate the part or flip it for placement on the opposite of the board for SMD layouts. When placing multiples of the same device type, *Tango-PCB Plus* automatically increments the part reference number as you go along. For example, if you start with a series of memory chips beginning with U10, *Tango-PCB Plus* will label the next one U11, the next U12, and so on. If there is already a device on the board with the next consecutive number, that number is skipped and the sequence continues with the next available higher number.

Both a ratsnest and force vectors are available to aid you in your placement. However, neither is available until all the parts in the netlist are on the board, which forces you to either guess at a best placement scheme or line the devices up in neat rows somewhere on the board, which takes less time, then use the move command in conjunction with either or both the ratsnest and force vectors for your layout.

Tracks are laid down by clicking the mouse button once and moving to the desired location. You may change layers while routing a track; vias are automatically placed when changing layers. *Tango-PCB Plus* supports eight copper layers, two of which are used strictly as power and ground planes. Track width cannot be changed on the fly, but you can edit the track width after routing. Connectivity is maintained when moving tracks. Parts can also be moved, rotated, or flipped while maintaining track connectivity. But it is only a direct path between the two pads, cutting across other tracks without concern, which leaves a bit of a mess that you will have to manually tidy up.

Tango-Route Plus is the autorouter companion to *Tango-PCB Plus,* and it too lists for $895. Purchased together, the price is $1,595. Like all Tango products, the autorouter runs alone rather than under the control of a main menu. To route a board, you load both the schematic netlist and the circuit board layout file from *Tango-PCB Plus.* The router then matches up the pin connections in the netlist to the device references in the PCB file and routes the tracks accordingly. Protected areas are defined in *Tango-PCB Plus* before routing.

The router has eight user-selectable routing methods, six of which are heuristic and two of which are Lee based. The Lee Maze router lets you specify up to 99 vias to complete a connection. All eight circuit board layers can be routed at the same time. There are two post-routing routines: one that minimizes the number of bends in a connection and one that minimizes the number of vias.

There are five grid sizes, ranging from 10 to 25 mils. Each grid comes with a specific track width and air gap clearance, which the user may change if desired. However, only one track width can be routed at a time. If you wish to route tracks of different widths, the nets involved must be highlighted and the routing pass done separately for each different width. On our benchmark layout using a 25-mil grid, the router was able to complete 100 percent of the tracks.

Tango-PCB Plus has no pan feature. Instead, you use the zoom function to move about the board, either to center your work or to move in or out. You may zoom in increments of two or use the infinitely variable window option. Zoom magnification ranges from filling the screen with a single pad to displaying the entire working area. Zoom is available from the keyboard during parts placement and routing. There is an undo command for both track placement and editing commands. Block functions include move, delete, copy, and save. A context-sensitive help screen is also available.

There is design rule checking in both the PCB layout and autorouter programs. Among the many things they check for is correct air-gap clearance and unrouted connections. All discrepancies are displayed in message boxes on the screen and saved to file.

Tango-PCB Plus generates a wide range of output files, including Gerber, N/C drill, DXF, and PostScript. There is also a full complement of masks, including an SMD paste mask, and several layers of silkscreen artwork.

Like all Tango products, the printer and plotter support is nothing but fantastic. However, after running the autorouter using the default values, you cannot use a printer hardcopy for real artwork. But if you pay attention to track width and spacing manually, you can use the artwork from a dot-matrix or laser printer to etch even relatively complex, double-sided circuit boards.

Tango-PCB Plus is a great package, if you do not mind paying the price and doing more manual labor than you

Table 16-1 Circuit Design Rating Summary

Schematic Capture: Tango-Schematic

Drawing	Editing	Library	Netlist Support	Ease of Use
good	excellent	excellent	poor	good

PCB Layout: Tango-PCB Plus

Place	Route	Library	Netlist Support	Ease of Use
poor	excellent	excellent	poor	good

should. It has everything you need to make professional multilayer boards, and the autorouter is among the best. But no support for other netlist formats or an automatic parts placement routine in a program that costs this much might have you looking for a better deal.

Chapter

17

CapFast, FutureNet-5, and PADS-PCB: À la Carte Software Combinations

The following three programs are sold as either standalone schematic capture or PCB layout programs. However, they are designed to be used together to form a complete circuit design package with a nearly seamless flow of operations from circuit capture to finished printed circuit board. The programs were selected because of their excellent, high-end performance.

The three programs discussed include *CapFast* from Phase Three Logic, *FutureNet-5* from Data I/O, and *PADS-PCB* from CAD Software. *CapFast* and *FutureNet-5* are schematic capture programs that interface intimately with PADS's PCB layout program. If *PADS-PCB* looks familiar, it's because CAD Software OEMs (sells) the program to Omation and Phase Three Logic, among others, for use in their *Schema-PCB* and *CapFast CF/PCB* PCB layout applications.

CapFast from Phase Three Logic

CapFast is an extremely comprehensive schematic capture program that can be used with virtually all circuit design software. The huge component library contains device parameters for every schematic capture want plus every post-schematic processing need, whether it be hardcopy, PCB layout, or circuit simulation. But at $799, it is not cheap.

Luckily for those of us who envy *CapFast*'s versatility but cannot afford it, Phase Three Logic offers a student version of *CapFast* that does everything the professional version does for just $299. The tradeoff is fewer parts per drawing (200 ICs or less) and decreasing drawing speed as the design grows more complex.

The professional version can use up to 16 MB of extended (not expanded) memory and needs at least 2 MB of RAM to run. The student version uses only 640K. Needed by both versions is an AT-compatible PC (286 or better) and a hard disk with about 14 MB of free space. A hardware protection key (dongle) is not required, and the software is not copy protected.

CapFast is moderately difficult to learn, but fairly easy to use. Commands are in hierarchy menus that may be as many as five levels deep. Choosing a root command usually has you select from a submenu, which may or may not lead to another menu. A completed command moves you back one menu so that it can be repeated; you can exit to the main menu at any time. Unless you have a three-button mouse, most commands are entered via the keyboard. Fortunately, macros that remember command sequences are supported.

Zoom has an infinitely variable range of 1000:1, with the highest magnification displaying about one-fourth of a logic gate. There is an adjustable autopan that moves the viewing screen by a user-defined amount whenever the cursor bumps against an edge. No help screen is provided, and you will probably have to keep the command reference manual under your nose for at least the learning period.

The enormous component library consists of about 9000 devices — 5000 of which are uniquely different. DeMorgan and IEEE equivalents exist for many of the devices, which, when added, push the total to over 10,000 parts. The part count includes 5000 TTL devices (1000 different types), 700 logic chips (CMOS, ECL, memory, etc.), 2400 diodes and transistors, 400 analog or passive devices, and 300 miscellaneous items like switches and connectors.

The library is divided into 17 menu selections. Included in the selections are libraries for SPICE, SUSIE, and Xilinx devices. New parts can be added using the library editor.

CapFast's strength is not in its sheer number of library devices, but in the information each device model contains. For example, each logic chip comes with 33 lines of data. Included are the typical device type, schematic reference, component value, and physical outline entries. Not so typical are instructions on which pins can be swapped during PCB layout, propagation times, and alternate package configurations. If that is not enough for your needs, you can add an unlimited number of data lines of your own using a nonformatted word processor or *CapFast*'s library editor.

Parts are called from the library by bringing up a parts list and clicking on the device for placement on the drawing. A listing of device parameters then appears. At this time you can specify the part value, schematic reference, or any of the many parameters the device has to offer. You can also specify which parameters to display (all if you wish), plus each one's text size and placement. But you may want to hold off on this until later, because placement editing, like rotation, can only be made after the part is in place, and your text could end up looking mighty strange. After the part is placed, you are returned to the parts list, where you may select another device for placement.

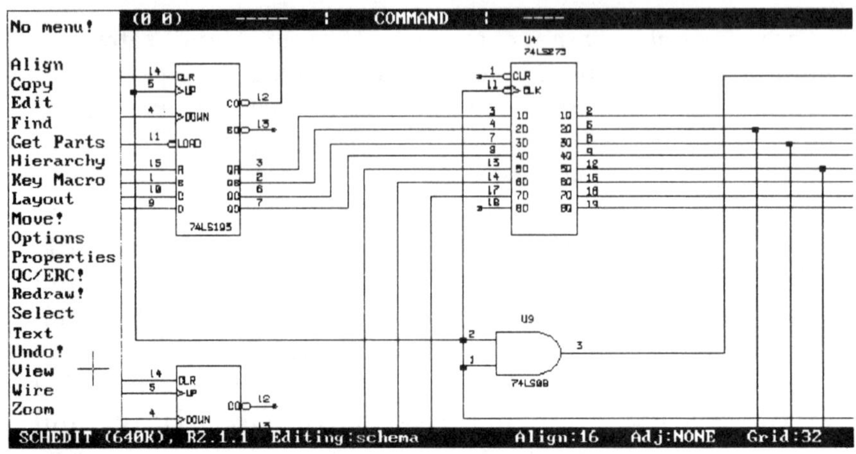

Figure 17-1 A *CapFast* schematic capture screen.

Wires can be drawn manually or automatically. Manually, you call up the Wire command and scoot the mouse from start to finish, clicking whenever you wish to change direction. Diagonals are permitted when ortho is turned off. For automatic wire routing, you can mark the wire's course by clicking the mouse on critical turning points and then activating the Wire command or by simply clicking on the destination and end points and letting *CapFast* find a route for you. Unfortunately, that route may be through the body of a component or on top of another wire. Whether wiring manually or automatically, a separate series of commands is needed for each wire placed.

You can also draw a wire manually, then have *CapFast* copy it as many times as you wish by simply tagging each wire's origin. This is most useful when working with buses or wiring memory arrays. But again, *CapFast* frequently places wires atop other wires unless the path is a direct route (no bends). Although the wire connections are electrically correct and can be used for netlist generation, you would never know it by looking at the schematic drawing.

However, each wire is individually labeled and can be identified if you choose to display the wire's name. This is called piping of wires as opposed to busing. Military and appliance manufacturers are heavy users of this technique. *CapFast* supports buses too.

CapFast has an excellent selection of schematic editing features, including move, delete, copy, rotate, and mirror. These tools are also available for block functions. When a part or block is moved, rubberband connections follow along, with *CapFast* doing its best to clean up the mess into an ortho drawing. But too often it misses the mark, making the electrically correct connection rather than the more visually correct separation of wires. Individual wires can be unstacked and rerouted if you can locate an elbow and move it to a new location. There is a find command, but no locate or jump. The undo command reverses your last command.

When it comes to generating netlists, *CapFast* is second to none. If the schematic models do not contain the information your PCB layout or circuit simulation program needs, you can add it. However, all netlists are generated in DOS using a *CapFast* conversion utility. While *CapFast* comes with an assortment of popular conversion formats, most notably *PADS-PCB* and *Schema III*, you may need to pay extra for the conversion program. Error checking for floating inputs, shorted outputs, and the like is provided.

Hardcopy output is decent, with support for 9- and 24-pin Epson printers plus Hewlett-Packard and Houston Instrument plotters. However, software printer and plotter control is limited to scaling (lets you reduce the drawing size) and page orientation.

As a stand-alone schematic capture program, *CapFast* is pricy and overkill. But for front end use in a complete circuit design package, it has no equal. It supports a huge device library and can interface with virtually every PCB layout and circuit simulation program around. If you work with a lot of different post-processing circuit design packages, this is the schematic capture program for you.

FutureNet-5 from Data I/O

FutureNet-5 is the industry standard for schematic capture software. It interfaces with virtually all PCB layout and circuit simulation software, comes with a large component library, and is chock-full of drawing and editing features. But at $895, it is among the most expensive.

The program requires only 640K of RAM to run, but will use up to 16 MB of LIM memory if available. A hard disk with 6.5 MB of free space is required, and the supplied hardware protection device (dongle) must be plugged into the parallel port before the program will run.

The program is moderately easy to learn and use. Commands are run from a two-level menu structure using either the keyboard or mouse. Many of the command screens are structured with Windows-like dialog boxes and command menus.

FutureNet-5 has both autopan and Windows-like scroll bars for moving around the drawing. Autopan abruptly jumps about half a screen when the cursor bumps against a screen edge. Zoom range is 10:1 and is available from the keyboard during parts and wire placement.

A context-sensitive help screen is available, and there is a macro recorder that remembers keyboard strokes and mouse clicks for later playback. Macros can be assigned to the top row function keys for fast execution.

The component library consists of about 4000 parts stored in 18 modules. The types of parts is quite diverse, ranging from TTL chips to microprocessors to Xilinx devices. The library modules have to be loaded individually into the drawing session after the program is started, but all can be active at the same time. Parts can be modified or added to any of the library modules using the library editor.

Parts can be transferred from the library to the drawing by selecting them from a scrolling parts menu or by typing their name at the command prompt. Components can be rotated and mirrored, but only after they are initially placed.

Device references are also added only after the device is placed on the drawing. The size and location of the annotation text is infinitely variable. There is an automatic numbering routine that sequentially numbers the parts in a raster scanning pattern, starting with the upper left corner and ending in the lower right corner.

Only ortho lines and wires can be drawn, not diagonals. *FutureNet-5* supports electrical wires, buses, and eight forms of lines, such as dotted, dashed, and so forth. Lines or wires are started by selecting the line command from the menu, placing the crosshairs at the beginning of the line, and clicking the mouse. Each successive click of the mouse changes the direction of the line and places the previous line segment on the drawing.

Drawing editing features are excellent, with support for move, delete, rotate, and copy. Blocks are also supported, and can be moved, erased, rotated, copied, and saved to file for use again. Connectivity is maintained when a part is moved, but you will have to approve the new wire placement one wire at a time. An undo command undoes commands in the reverse order they were entered; the redo command reverses the undo command by redoing actions that have been undone. Undo and redo are available only if LIM memory is present.

The program outputs the schematic in the popular FutureNet netlist standard. This netlist format is supported by a large number of PCB layout programs. Netlist

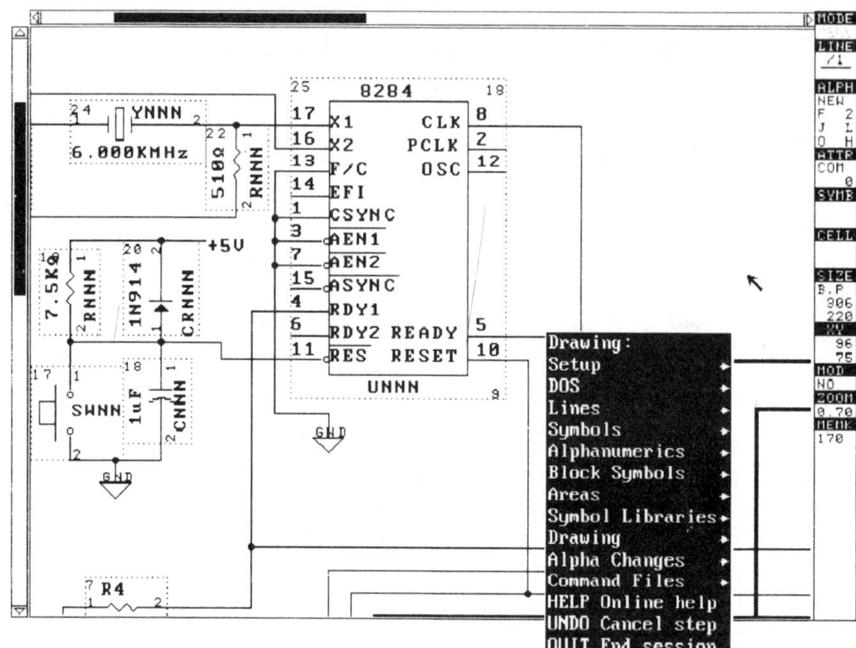

Figure 17-2 A *FutureNet-5* schematic capture screen.

generation is done in DOS. A DOS-based error checking utility checks for floating inputs, shorted outputs, and broken wires.

FutureNet-5 supports almost 200 dot-matrix and laser printers. If the drawing is larger than the page, it is printed on multiple pages that have to be taped together. Plotters are not supported.

FutureNet-5 is an excellent schematic capture program to which other schematic programs are compared. The component library is large and the editing features are plentiful. Unfortunately, it is not the easiest program to use and its price is very high.

PADS-PCB from CAD Software, Inc.

PADS-PCB is one of the best PCB layout programs around. It supports automatic parts placement, three autorouters, and an array of editing tools that are second to none. But get your pocketbook ready, because its $975 base price is just the beginning. A full-blown package can cost $2,300 and more.

The program can support up to 8 MB of LIM memory if available, but requires only 640K to run. A hard disk is needed, and the supplied hardware protection key must be plugged into the parallel port before the program will run.

Access to the program is via a common main menu that lets you access other PADS-related programs, like schematic capture and the library editor. Going from PCB layout to artwork generation is not exactly seamless, but the main menu keeps you from having to run each program from a DOS prompt.

The program is easy enough to learn but more difficult to use because the commands are stacked in a hierarchic structure that has you accessing layer after layer when searching for a command. What's worse is that you have to follow the menus backwards one by one when exiting. Fortunately, the process can be expedited by pressing the first letter of the command from the keyboard, saving you having to race the mouse around the desktop.

Zoom and pan are manual, not automatic. However, both can be used during parts placement and routing. Zoom range is 64:1. A help screen is not provided.

The library contains outlines for 159 devices, including several SMDs, that are electrically linked with 2160 components, mostly logic chips. Parts can be added to the library using a library editor that is accessible from the main menu.

PADS-PCB has an excellent automatic parts placement program ($350 extra) that can even properly place

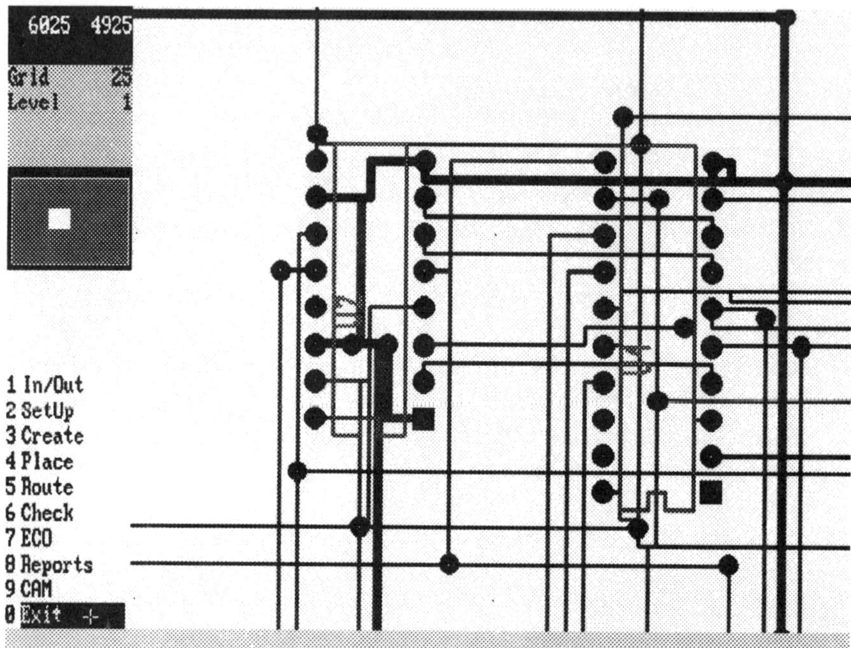

Figure 17-3 A *PADS-PCB PCB* layout screen.

decoupling capacitors. But before devices can be extracted from the netlist for automatic placement, you have to define both a board outline and a parts matrix. Maximum board size is 32 × 32 inches.

The matrix is not sized to the board so that you can place sections of the board according to device type, such as assigning one area to memory chips only. While this is a great asset, it takes some planning to avoid getting the layout too cluttered or too spread out. A matrix is not required for manual placement.

Parts can be rotated or flipped to the solder side of the board for SMD layout after auto placement. Any part can be glued in place and the automatic placement run again for improved placement.

There are three autorouter options, all at extra cost. The standard autorouter ($750) uses a mix of heuristic and Lee algorithms that are user selectable. The user may choose from 15 routing modes and run them either separately or as a batch. The autorouter runs through the list beginning with power supply tracks and ending in via optimization. However, you can autoroute only two layers and one track width at a time. Each different layer pair or new track width requires a separate route.

Autorouting can be confined to a net or specific area of the board and may be interrupted at any time for manual intervention. The router grid is variable between 1 and 800 mils and can be changed on the fly.

A rip-up router that can route 12 layers simultaneously sells for $3,500 and a shove-aside router goes for $1,495.

For manual routing, the tracks are started by clicking on a pad and moving the pointer to the connecting pad. A ratsnest, which can hide power supply nets, is always visible. Changing the track width for necking down can be done on the fly. Curved tracks are not supported. The program supports 30 copper layers, and each can have both a power and a ground plane.

PADS-PCB is one of the very few PCB layout programs that lets you change package types (e.g., from DIP to SMD) after the parts are placed. It is also one of the few that lets you do gate and pin swapping.

Editing tools include move, rotate, delete, mirror, and copy for both individual tracks or parts and blocks of components. Blocks can also be saved to file for use again. Track connectivity is maintained when moving a device or block.

The design rule checker checks for air-gap violations as well as floating inputs and shorted outputs. There is also a density map that displays the concentration of tracks on the board and a histogram that plots track density on a screen chart. Both can be used to open up congested areas

Table 17-1 Circuit Design Rating Summary

Schematic Capture: CapFast

Drawing	Editing	Library	Netlist Support	Ease of Use
excellent	excellent	excellent	excellent	fair

Schematic Capture: FutureNet-5

Drawing	Editing	Library	Netlist Support	Ease of Use
excellent	excellent	excellent	excellent	good

PCB Layout: PADS-PCB

Place	Route	Library	Netlist Support	Ease of Use
excellent	excellent	good	excellent	fair

or used in conjunction with the optional thermal analysis programs (prices start at $1,995) for cooling analysis.

The program can produce a full range of output netlists and files — at extra cost, of course. The Gerber and N/C drill utilities are $300 each, and the DXF file generator is $495. The only thing you do not have to pay extra for are the layer masks, which include all 30 copper layers, solder and SMD paste masks, and top and bottom layer silkscreen artwork.

A dot-matrix or laser printer can be used to produce artwork good enough for limited production and prototype work. A good assortment of Hewlett-Packard and Houston Instruments pen plotters are also supported. Artwork may be sized to fit the page, scaled to a model size, or printed in actual size.

PADS-PCB is an excellent PCB layout package that has automatic part placement, support for up to 30 layers of copper, and permits package, pin, and gate swapping. There are three autorouter options, and the program can be used with nearly all schematic capture programs using one of the many optional netlist conversion utilities. The bad news is its price, which can add up to a tidy sum by the time you buy all the needed options.

Chapter

18

Hardware Requirements

With but a couple of exceptions, the hardware requirements for a schematic capture program are the same as for a drawing or a CAD program. Some of these requirements were touched on briefly in Chapters 2 and 5. But here is the real story. Use the following guidelines to select the PC system that is correct for your needs. Buying smart means not buying more hardware than you can use.

Graphics Display

Topping the list of hardware needs is high-resolution screen graphics. The most popular video mode in use today is IBM's Video Graphics Array (VGA). VGA originally came packaged only with the IBM PS/2 computer, but has since been cloned for use with all IBM compatibles and is the cheapest video option available for the PC. With a screen resolution of 640 × 480 pixels in 16 colors, the VGA screen is sharp and colorful and is supported by all schematic capture software — making it the ideal video display for entry-level schematic capture.

For complex design work, however, you need to go beyond VGA. Among the new video modes available for printed circuit board design work are super-VGA, 8514/A, and high-resolution graphics coprocessor boards. These boards offer screen resolutions that range from 800 × 600 up to 1600 × 1200.

Unlike VGA, however, there is no single video standard beyond VGA that works with all schematic capture software. The closest anyone has come is the Video Electronics Standards Association (VESA), which has a super-VGA standard for both 800 × 600 and 1024 × 768. Although hardware support for VESA is widespread, software support is slow in coming, which means that until a single standard is adopted by all, you have to match your hardware and software one on one.

Super-VGA

The hot ticket today is super-VGA because it lets you display high-resolution graphics at a reasonable price. Super-VGA is found in combination with most VGA boards and costs only a few dollars more than VGA alone. Virtually all VGA boards have an 800 × 600 mode, and about 25 percent also support a 1024 × 768 mode.

However, you can't depend on widespread software support for super-VGA. Because of the sheer number of super-VGA boards on the market, it takes some real sleuth work to find a super-VGA board that is supported by your schematic capture software. Your best bets are boards built around a Tseng, Paradise, or Headland Video-Seven graphics chip.

Displaying a super-VGA mode also requires a special monitor, and each video mode has monitor requirements

Figure 18-1 Super-VGA boards have better screen resolution than standard VGA, like this one from Micro Express, which has a top resolution of 1600 × 1024.

that are different from all other video modes. The trick is to match your super-VGA mode to a compatible super-VGA monitor. See "Monitors" in this chapter for more details.

But in exchange for low price, expect tradeoffs in screen speed and image quality — especially at 1024 × 768. At that resolution, your super-VGA board may turn super slow, taking several seconds to refresh a screen. Although there are ways to speed up super-VGA operations, like relocating the video BIOS to shadow RAM or using special VRAM memory chips, they actually do little to enhance screen performance.

Moreover, most super-VGA boards use *interlaced scanning* at 1024 × 768 (VGA and super-VGA 800 × 600 are noninterlaced). This means the electron gun inside the monitor takes two passes to create an image instead of one, which causes screen flicker and a reduction in image sharpness.

How do you solve these performance and quality problems? Easy: Spend more money.

Graphics Coprocessor Boards

Faster speeds can be realized by using boards with special graphics coprocessors. A graphics coprocessor is a lot like a math coprocessor in that it unloads the video information from the PC and processes it independently of the CPU. Speed increases of up to 25 times are possible, depending on the program and graphics coprocessor type.

8514/A The cheapest graphics coprocessor board is the 8514/A video adapter. First introduced by IBM as a high-resolution extension of VGA, the 8514/A board displays 1024 × 768 interlaced graphics and only costs about $200 more than a super-VGA board with the same resolution. Although 8514/A screen quality is not quite as sharp as the more expensive graphics coprocessor products because of its interlaced scanning, it is an IBM standard and therefore has greater software support than super-VGA boards. However, support for 8514/A is not as widespread as VGA, and you should not expect your schematic capture software to blindly support it.

Many software makers are hesitant to provide 8514/A support because of IBM's marketing decision to limit the board for use with PS/2 computers that have Micro Channel Architecture (MCA) only. IBM has no generic version of the 8514/A, making it difficult for software programmers to justify the time spent on writing special hardware drivers for a limited number of users. Although the 8514/A has since been cloned by other board makers for use with IBM compatibles in a standard AT slot, it may be too little too late. The bottom line is, check your schematic capture software for 8514/A support before buying 8514/A hardware.

Also be aware that the 8514/A board cannot be used alone. It is a 1024 × 768 graphics controller only, and as such it can only display programs that write to the 8514/A screen. Programs that use VGA — including DOS — cannot be processed by the 8514/A board, which means you need two monitors: one for DOS (so that you can control the PC) and one for the schematic capture program.

However, 8514/A has a pass-through feature that lets you display both VGA and 8514/A on an 8514/A monitor (not at the same time, of course). The secret is in the 8514/A monitor, which is a dual-frequency monitor that can sense the difference between the VGA and 8514/A modes and adjust itself to display each properly. When the 8514/A board is not in the 1024 × 768 mode, it allows the VGA signal to pass through the board to the monitor. But as soon as an 8514/A application is activated, the VGA signal is suppressed and the 8514/A video is sent to the monitor. The normal scenario is to power up the PC under VGA, then load the schematic capture software, which causes the monitor to switch over to 1024 × 768 resolution. Exiting the schematic capture program automatically switches the monitor back to VGA format and the DOS prompt.

Although the VGA pass-through feature lets you get by with just one monitor, you still need two video boards: one for VGA and one for 8514/A. Sometimes the VGA electronics are built into the 8514/A board, eliminating the need for a second board, but mostly a separate VGA board is required, which means you will have to buy a VGA board in addition to the 8514/A if your PC doesn't already have one. The 8514/A board is connected to the VGA card via a 22-pin features connector located on the top of the VGA board using a patch cable. In the PS/2, the 8514/A board plugs into a special video slot that connects the board to the motherboard's VGA circuitry for pass-through.

IBM has recently announced the release of XGA, which is nothing more than 8514/A with a built-in VGA controller. This eliminates the need for two cards. However, XGA is currently available only in IBM's PS/2 systems. And while XGA *seems* to be fully compatible with 8514/A software, there are some reported cases of software incompatibility.

TIGA For top-of-the-line speed and image quality, it's hard to beat boards built around Texas Instruments' TMS340 graphics processor chips, the 34010 and the 34020. TMS340-based graphics boards have been available for longer than either VGA or 8514/A, and consequently have a more mature technology and a substantial software following (albeit not as large as VGA). However, they are the most expensive of the trio.

Unlike super-VGA and 8514/A boards, which peak at 1024 × 768, TMS340-based boards begin with 1024 × 768 noninterlaced graphics and improve from there. Many have 1280 × 1024 screens, and a few offer 1600 × 1200.

Like the 8514/A, the TMS340 chips cannot process VGA signals and must be used with dual monitors or provide a VGA pass-through with a special monitor. However, a large percentage of TMS340 boards include VGA circuitry either as a part of the board or as an optional add-on module that plugs into the board, thus eliminating the need for a separate VGA board. But it does not eliminate the need for a special monitor that operates at VGA scan rates in addition to high-resolution TMS340 frequencies. TI recently announced a TMS340 VGA interface chip that allows the 34010 and 34020 to process software applications written for VGA, 8514/A, and pre-VGA standards (MDA, CGA, and EGA) and display them on a standard fixed-frequency, high-resolution monitor — cutting monitor cost considerably. Boards using the new TMS340 VGA interface chip should appear soon.

The strength of TMS340-based boards lies in its software-driven interface. Instead of writing to the hardware registers of these two chips, as you do with VGA, you write a software program in a compiled TI language that draws the screen elements for you. Using this method allows Texas Instruments to make improvements in the TMS340 hardware without affecting software compatibility, and it allows the software application to use any and all high-resolution modes supported by the TMS340 board via a single video driver. You do not need individual drivers for each video mode.

Several TMS340 interface standards have been developed over the years, but the two most popular are DGIS (Direct Graphics Interface Standard) and TIGA (Texas Instruments Graphics Architecture), either or both of which can be found in many schematic capture programs.

Monitors

The monitor is the most important part of the video system. No matter how good the video controller board may be, the final screen image is only as good as the monitor itself. Buy the wrong monitor and you pay for it with eyestrain and fatigue. On the other hand, there is no point in spending money for features you do not need or cannot use.

Screen colors are important to circuit design because the software makes extensive use of color if it is available. Copper layers are a good example of where color is imperative in identifying the level you are working on. Using colors to identify areas of a circuit also makes it easier to edit the drawing. Users that have monochrome displays, like Hercules Graphics users, will find that their higher-resolution screens excel in sharpness but lack the information a color screen provides. Generally, the extra dollars spent on a color display is money well spent.

If possible, you should always select a monitor by comparing your monitor choices side by side with the video board you plan on using. Some monitors will adjust automatically to the board, while others will require manual adjustment to get the picture size right. Some will not align properly at all — indicating either a bad match between board and monitor or a defective unit. Here are the four factors to look for when shopping for a monitor: compatibility, scanning method, dot pitch, and screen size.

Compatibility

The most important thing is to match the monitor to the video board. A VGA board requires a VGA monitor, and an 8514/A requires an 8514/A-compatible monitor. You cannot use a VGA monitor for 800 × 600 or 8514/A.

There are three types of video monitors. The first is a fixed-frequency monitor like the IBM 8514 that works with one video mode only, generally VGA. Next in price and compatibility are dual-frequency monitors that work with two video modes, such as VGA and super-VGA 800 × 600. The most compatible and most expensive is the multiscan monitor, which works with a range of video modes.

Dual-frequency monitors are fixed-frequency monitors that handle two sweep rates. The 8514/A monitor is an example of a dual-frequency monitor, with support for both 640 × 480 noninterlaced and 1024 × 768 interlaced screens. However, the 8514/A monitor cannot display 800 × 600 super-VGA graphics. For that you need a different dual-frequency monitor, like the NEC 2A, or a multiscan monitor. Multiscan monitors use a variable-frequency sweep that spans a specified range. Any video mode that falls within the limits of the sweep range can be displayed, regardless of whether it's VGA, 800 × 600, or 1024 × 768.

The decision to buy a dual-frequency or multiscan monitor depends on how many video modes you intend to support and how much you are willing to spend. Dual-frequency monitors cost less than multiscans, so if you work in one super-VGA mode exclusively, a dual-frequency monitor can save you money. However, if your schematic capture software supports three or more beyond-VGA modes, it pays to go with a multiscan monitor because it lets you upgrade to better super-VGA resolutions as they are invented, saving you the cost of a new monitor. And while you may intend to use a monitor at a beyond-VGA resolution, some offices also need the ability to move the monitor among other PCs with widely varying video controller types, a job only a multiscan monitor can do.

However, not all multiscan monitors support all video modes. There is a tradeoff between the number of video modes the multiscan monitor can display and price, and many multiscan monitors are purposely limited in range to reduce cost. The determining factor is the capture range, or the span of horizontal and vertical scan frequencies the multiscan monitor can lock onto. The wider the capture range, the more video modes supported and the more expensive the multiscan monitor. A number of the higher-priced multiscan monitors also handle digital input, which means they work with VGA, EGA, CGA, monochrome, and Hercules Graphics boards.

To ensure that the video board and monitor are compatible, compare the video scan frequencies listed in Table 18-1 to the monitor specs.

Scanning Method

Recent studies show that the eyestrain and fatigue you experience from staring at the screen are in large part caused by screen flicker. Flicker occurs when the screen image

Table 18-1 Video Mode Scan Frequencies

Video Mode	Horizontal Rate (kHz)	Vertical Rate (Hz)
CGA	15.75	60
MDA	18.0	60
EGA	21.8	60
VGA	31.5	70
VESA	35.2	56
8514/A	31.8	43

fades before the PC has time to refresh the image. The effect is similar to an old-time nickelodeon show, but at a rate so fast you may not see it — but your eyes feel it.

The best way to reduce flicker is to increase the number of times per second the screen is redrawn. In technical terms, this is known as increasing the refresh rate. Video modes that use interlaced scanning are particularly bad because their refresh rate is a slow 43 Hz, a rate so slow that the image flicker is actually noticeable by many users. Most noninterlaced video modes, on the other hand, have a refresh rate of 60 Hz, which does not flicker for most users. Although 60 Hz is fast enough that your brain cannot tell the screen is flickering, your eyes sure can.

A new standard proposed by the Video Electronics Standards Association (VESA) increases the refresh rate from the current 60-Hz standard to 72 Hz. This eye-saving technique offers the added bonus of slightly crisper text and graphics. However, only top-of-the-line multiscan monitors (noninterlacing 1024 × 768 models) can support the new refresh rate. But help may be on the way. A number of European countries are demanding 72 Hz as a minimum standard, which should increase manufacturing volumes and lower the cost of this healthy technology.

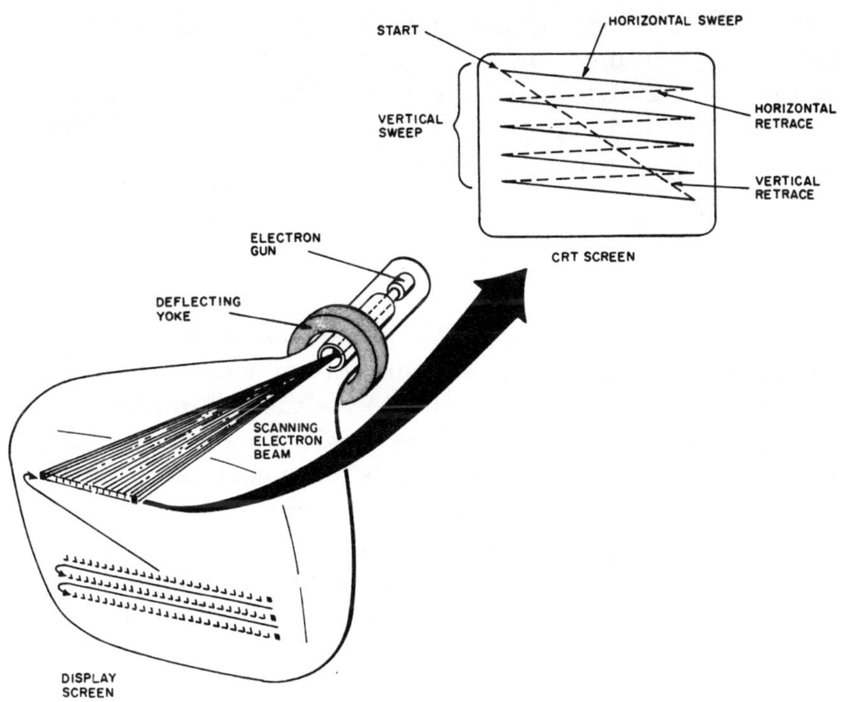

Figure 18-2 Most super-VGA boards use interlaced scanning for resolutions of 1024 × 768 and beyond, an inexpensive solution that can lead to flicker and eye fatigue.

Dot Pitch

Next to resolution, dot pitch is the most important factor determining the apparent sharpness of an image. The dot pitch is the distance between the tiny phosphor dots on the screen's inner surface. The smaller the dot pitch, the sharper the image.

The most co mmon dot pitches are 0.31 mm and 0.28 mm. On a 14-inch monitor, a 0.31 mm dot pitch is adequate for running at 800 × 600, but if you want to up the ante to 1024 × 768, you need a 0.28 mm dot pitch or better. On a 16-inch monitor running at 1024 × 768, greater screen area makes a 0.31 mm dot pitch acceptable. For 1280 × 1024 and beyond, only 0.28 mm dot pitch or better on a 16-inch screen is acceptable.

Screen Size

Image detail is determined by the screen size. As screen resolution increases, the amount of information on the screen also increases. To be able to see the information, the screen size has to increase. The minimum reco mmended size for schematic capture work is 14 inches, and that is only for resolutions up to 800 × 600. For 1024 × 768 and above, nothing less than a 16-inch screen is reco mmended. And since you are buying a 16-inch monitor anyway, pay the extra $100 to $200 for noninterlaced scanning and treat yourself to a flicker-free environment. A tilt-and-swivel base lets you adjust the screen for comfortable viewing and is essential for minimizing glare from ambient light sources.

Monitor sizes beyond 16 inches can get monstrous, with some screens growing as large as 35 inches. However, you should draw the line at 19 inches (sometimes advertised as 21 inches), because it is about all the monitor you have room for on a standard desktop. Moreover, many monitors beyond 19 inches require special video controller boards, which may or may not be compatible with your schematic capture or PCB layout software.

Choosing the Right PC

There was a time when any piece of software written for the IBM PC would run on any version of the PC or any compatible. This no longer holds true, with more and more software being written for the 80386 PC. Pushing that technology is not the business class programs, but CAD software, like circuit design programs. So in addition to worrying about CPU speed, you must also concern yourself with CPU type.

CPU Type

The history of the PC begins in late 1981, with IBM's announcement of the IBM PC. This was an 8088-based machine with just 64K of RAM and a 4.77-MHz clock speed. Within a matter of months a flurry of software deluged the market, all in support of the computer marvel. Of course, most of that software has long been forgotten, but it set in motion a trend that would not change for nearly a decade.

The problem arose with Intel's announcement of the 80286 chip, which was faster and could access more memory — and which IBM stated would be used in their new personal computer, the IBM PC/AT. While the new chip did deliver better performance, very few progra mmers were willing to relinquish their share of an already large 8088 market. Consequently, little software progress was made, and programs continued to be written in support of the 8088.

In 1989, Intel began shipping its 80386 chip. Again, Intel upped the ante on extended memory support and added several extremely attractive progra mming features. This time it worked, and 80386-specific application began to appear, most notable is Microsoft's 386 version of *Windows*.

Although there is a trend toward 386-specific software in the circuit design co mmunity, most of the software sold is written for the 8088- and 80286-based PC/AT. The exceptions are circuit simulation programs, like SUSIE 6.0, which require a 386 PC just to get started. If you need a 386 PC, but live on a tight budget, consider a 386SX machine.

Speed

The speed of the PC is critical only when you work with large designs and frequent screen refreshes. Nonetheless, you should never consider anything less than 12 MHz, with 16 MHz being a better choice.

However, when you start looking at the 16-MHz market, you will notice that the prices of 386SX PCs are only slightly higher than that of a comparable PC/AT. And because 80286 software will run on a 386SX as well as on a 386 PC, there's little to think about. The 16-MHz 386SX is the better buy, because it gives you room to grow as the new software becomes available.

High-end users may want to consider 25-MHz and 33-MHz 386 systems as their primary tool, as much for the added expanded memory capacity as room for a larger hard disk. While cheaper than they were just a year ago, they still approach the $10,000 mark by the time everything is added. There is no advantage to buying a 486 system at this time.

Memory

Most schematic capture and PCB layout programs are extremely memory hungry. The rationale behind using expanded memory is simple: Keep a large file entirely in RAM instead of on disk, and the speed increases dramatically. Some programs cannot even load unless expanded memory is installed. The average need is between 2 and 8 MB, depending on the program.

Most programs use LIM 4.0 Expanded Memory Software (EMS). But there are a few that use extended memory instead. The difference is that extended memory is what is found in the PC beyond 1 MB. It is raw memory just there for the taking. EMS memory, on the other hand, is extended memory that is under the control of an expanded memory manager program. EMS memory is easier for the circuit design software to use because its specifications are well defined. When writing a program to use extended memory, the progra mmer must invent his own memory manager and add it to the other software. LIM 4.0 is the established EMS standard.

Hard Disk

Schematic capture and PCB layout programs also require a lot of hard disk space. As a rule, plan on spending at least 3 MB of disk space for the program itself, with some programs like *CapFast* demanding as much as 15 MB. Then you have to factor in another 5 to 500K per drawing, plus the netlists it will generate. 40 MB should be the smallest hard disk you should consider, with 120 MB being none too big.

Math Coprocessor

Because most circuit design operations are math intensive, the program generally benefits greatly when a math coprocessor chip is installed in the PC. Typically, a math coprocessor can increase program operations speed by at least two under normal conditions and by as much as 10-fold when doing a screen redraw. But check the software before going to the several-hundred-dollar expense of installing a math coprocessor, because there are several programs out there that do not use them.

Appendix

Schematic Capture Features

	CapFast	DC/CAD	EE Designer III	FutureNet-5
Standard Features				
Price	$799	$495	$995	$895
Integrated PCB software	N	Y	Y	N
Hardware protection key	N	N	Y	Y
Copy protected	N	N	N	N
Hard disk required	Y	Y	Y	Y
Help screen	N	Y	N	Y
Drawing Features				
Ortho disable	Y	Y	Y	N
Diagonals in ortho mode	N	Y	N	N
Pull down menus	Y	N	Y	Y
Auto annotate	Y	N	Y	Y
Supports buses	Y	Y	Y	Y
Repeat object placement	Y	N	Y	N
Zoom range	1000:1	unlimited	100:1	10:1
Auto pan	Y	Y	N	Y
Warns/removes duplicate objects	Y	N	Y	
Jump	N	Y	N	Y
Find/search	Y	Y	N	
Editing Features				
Placement				
Rotate	Y	Y	Y	Y
Mirror	Y	Y	Y	Y
Find	Y	Y	Y	Y
Undo	Y	Y	Y	Y
Rubberband	Y	Y	Y	Y
Fixup	Y	N	N	Y
Block				
Move	Y	Y	Y	Y
Hide	N	N	N	N
Delete	Y	Y	Y	Y
Save	Y	Y	Y	Y

	CapFast	DC/CAD	EE Designer III	FutureNet-5
Part reference				
Move	Y	Y	Y	Y
Hide	Y	Y	N	N
Delete	Y	Y	Y	Y
Adjustable font size	Y	Y	Y	Y
Library				
Screen directory	Y	N	N	Y
DeMorgan	Y	N	N	Y
Number of modules	17	11	1	18
Maximum number of active modules	unlimited	1	1	18
Printer Output				
Fit to page	Y	Y	Y	N
Border off	Y	N	N	Y
Title box off	Y	N	N	Y
Software control	N	N	Y	N
Plotter Output				
Fit to page	Y	Y	Y	na
Maximum page size	E	E	E	na
Software control	N	Y	Y	na

	HiWIRE II	ISIS Supersketch	OrCAD/STD III	ProCAD
Standard Features				
Price	$995	$149/$199	$495	$695/$795
Integrated PCB software	Y	N	N	Y
Hardware protection key	Y	N	N	N
Copy protected	N	N	N	Y
Hard disk required	Y	N	N	Y
Help screen	N	N	N	Y
Drawing Features				
Ortho disable	N	Y	Y	Y
Diagonals in ortho mode	N	Y	Y	Y
Pull down menus	Y	Y	Y	N
Auto annotate	N	Y	(1)	Y
Supports buses	Y	Y	Y	Y
Repeat object placement	N	Y	Y	Y
Zoom range	128:1	20:1	20:1	unlimited
Auto pan	Y	Y	Y	Y
Warns/removes duplicate objects	Y	Y		
Jump	N	Y	Y	N
Find/search	Y	Y	Y	Y
Editing Features				
Placement				
Rotate	Y	Y	Y	Y
Mirror	Y	Y	Y	Y
Find	Y	Y	Y	Y
Undo	Y	Y	N	Y
Rubberband	Y	Y	Y	Y
Fixup	N	Y	Y	N

	HiWIRE II	ISIS Supersketch	OrCAD/STD III	ProCAD
Block				
Move	Y	Y	Y	Y
Hide	Y	N	N	Y
Delete	Y	Y	Y	Y
Save	Y	Y	Y	Y
Part reference				
Move	N	Y	Y	Y
Hide	N	Y	Y	Y
Delete	N	Y	Y	Y
Adjustable font size	N	Y	N	Y
Library				
Screen directory	Y	Y	Y	Y
DeMorgan	N	Y	Y	N
Number of modules	7	7	11	9
Maximum number of active modules	1	1	10	1
Printer Output				
Fit to page	N	Y	Y	Y
Border off	N	N	N	N
Title box off	N	N	Y	N
Software control	N	N	N	Y
Plotter Output				
Fit to page	Y	Y	Y	Y
Maximum page size	E	A0	E	E
Software control	Y	N	N	Y

	Protel-Schematic	Schema III	Schema-Quik	SuperCAD
Standard Features				
Price	$495	$495	$99	$99
Integrated PCB software	N	N	N	N
Hardware protection key	Y	N	N	N
Copy protected	N	N	N	N
Hard disk required	N	N	N	N
Help screen	N	Y	Y	Y
Drawing Features				
Ortho disable	Y	Y	Y	Y
Diagonals in ortho mode	Y	N	N	Y
Pull down menus	Y	Y	Y	Y
Auto annotate	Y	N	N	Y
Supports buses	Y	N	N	Y
Repeat object placement	Y	N	Y	Y
Zoom range	20:1	8:1	8:1	two levels
Auto pan	Y	Y	Y	N
Warns/removes duplicate objects	Y	N	Y	N
Jump	Y	N	N	N
Find/search	N	Y	Y	Y

	Protel-Schematic	Schema III	Schema-Quik	SuperCAD
Editing Features				
Placement				
Rotate	Y	Y	Y	Y
Mirror	Y	N	N	N
Find	N	Y	Y	Y
Undo	N	Y	Y	Y
Rubberband	Y	Y	Y	Y
Fixup	N	Y	Y	N
Block				
Move	Y	Y	Y	Y
Hide	Y	N	N	N
Delete	Y	Y	Y	Y
Save	Y	N	N	Y
Part reference				
Move	N	N	N	N
Hide	Y	N	N	N
Delete	N	N	N	N
Adjustable font size	Y	N	N	N
Library				
Screen directory	Y	N	Y	Y
DeMorgan	N	Y	Y	Y
Number of modules	14	9	1	11
Maximum number of active modules	10	9	10	11
Printer Output				
Fit to page	Y	N	N	N
Border off	Y	Y	Y	Y
Title box off	Y	Y	Y	Y
Software control	Y	N	N	N
Plotter Output				
Fit to page	Y	Y	Y	na
Maximum page size	E	E	E	na
Software control	Y	Y	Y	na

	Tango-Schematic
Standard Features	
Price	$495
Integrated PCB software	N
Hardware protection key	Y
Copy protected	N
Hard disk required	Y
Help screen	Y
Drawing Features	
Ortho disable	Y
Diagonals in ortho mode	Y
Pull down menus	Y
Auto annotate	Y

	Tango-Schematic
Supports buses	Y
Repeat object placement	Y
Zoom range	unlimited
Auto pan	N
Warns/removes duplicate objects	Y
Jump	N
Find/search	Y
Editing Features	
Placement	
Rotate	Y
Mirror	Y
Find	Y
Undo	Y
Rubberband	Y
Fixup	Y
Block	
Move	Y
Hide	N
Delete	Y
Save	Y
Part reference	
Move	Y
Hide	Y
Delete	N
Adjustable font size	N
Library	
Screen directory	Y
DeMorgan	Y
Number of modules	23
Maximum number of active modules	10
Printer Output	
Fit to page	Y
Border off	Y
Title box off	Y
Software control	Y
Plotter Output	
Fit to page	Y
Maximum page size	E
Software control	Y

1) Via an external DOS utility

PCB Layout Features

	Labcenter DC/CAD	EE Designer III	HiWIRE II	PCB II
Standard Features				
Price	$495	$995	$995	$149
Integrated schematic capture software	Y	Y	Y	N
Hardware protection key	N	Y	Y	N
Copy protected	N	N	N	N
Supports EMS RAM	3MB	8MB	32MB(1)	N
Hard disk required	Y	Y	Y	N
Help screen	Y	N	N	N
Maximum board size (inches)	32 x 32	32 x 32	64 x 64	30 x 30
Netlist Features				
Front annotate	Y	Y	Y	N
Back annotate	Y	Y	N	N
Bill of materials	Y	Y	Y	Y
Parts mask	Y	Y	Y	N
Solder mask	Y	Y	Y	Y
Silkscreen masks	Y	Y	Y	Y
Component Placement				
Auto place	Y	Y	N	N
Glue in place	Y	Y	N	N
Protect zones	Y	Y	N	N
Decouple caps	N	N	N	N
Opposite side	Y	Y	Y	Y
Add on fly	Y	Y	N	N
Ratsnest	N	Y	N	N
Vector force	N	N	N	N
Hide paths	N	N	N	N
Auto pan	Y	N	Y	Y
Zoom range	unlimited	100:1	128:1	50:1
Routing Features				
Autoroute				
Nonstop	Y	Y	Y	N
Interactive	Y	Y	Y	N
Pad to pad	N	Y	N	N
Mix widths	N	Y	N	N
Lee	N	Y	Y	N
Heuristic	N	N	N	N
Rip-up router	Y	N	Y	N
Number of copper layers	64	12	64	2
Ortho off	Y	Y	Y	Y
Curved tracks	N	Y	N	N
Auto vias	Y	Y	Y	Y
Ground planes	Y	Y	Y	Y
Protect zones	Y	Y	N	N
Neck down	N	Y	N	N

	Labcenter DC/CAD	EE Designer III	HiWIRE II	PCB II
Editing Features General				
Find/search	Y	N	Y	Y
Jump	Y	N	N	Y
Change colors	Y	Y	Y	Y
Macros	Y	Y	N	N Place
Move	Y	Y	Y	Y
Rotate	Y	Y	Y	Y
Opposite side	Y	Y	Y	Y
Card outline	Y	Y	Y	Y
Change package	N	Y	N	N
Change pad	Y	Y	Y	Y
Pair swapping	Y	Y	N	N
Gate swapping	N	Y	N	N
Pin swapping	N	Y	N	N Block
Move	Y	Y	Y	Y
Delete	Y	Y	Y	Y
Copy	Y	N	Y	Y
Save	Y	Y	Y	Y Route
Delete track	Y	Y	Y	Y
Change width	Y	Y	Y	Y
Chg via size	Y	Y	Y	Y
Via optimize	Y	Y	Y	N
Rubberband	Y	Y	Y	Y
Density map	N	N	N	N
Histogram	N	N	N	N
Design Rule Check				
Counts number of vias	N	Y	Y	N
Counts number of nets	N	Y	N	N
Reports unconnected nets	Y	Y	Y	N
Reports broken tracks	N	Y	N	N
Reports air-gap violations	Y	Y	Y	N
Output Formats				
HPGL plotter	Y	Y	Y	Y
HI plotter	Y	Y	Y	Y
PostScript	N	N	Y	Y
Dot-matrix printer	Y	Y	Y	Y
Gerber	Y	Y	Y	Y
N/C Drill	Y	Y	Y	Y
DXF	Y	Y	Y	N

	OrCAD/PCB II	PADS-PCB	PCBoards	ProCAD
Standard Features				
Price	$1,495	$975	$99	$695/$795
Integrated schematic capture software	N	N	N	Y
Hardware protection key	N	Y	N	N
Copy protected	N	N	N	Y
Supports EMS RAM	N	8MB	N	8MB

	OrCAD/PCB II	PADS-PCB	PCBoards	ProCAD
Hard disk required	N	Y	N	Y
Help screen	N	N	Y	Y
Maximum board size (inches)	32 x 32	32 x 32	6 x 13	64 x 64
Netlist Features				
Front annotate	Y	Y	N	Y
Back annotate	Y	Y	N	Y
Bill of materials	Y	Y	N	Y
Parts mask	Y	Y	N	Y
Solder mask	Y	Y	N	Y
Silkscreen masks	Y	Y	N	Y
Component Placement				
Auto place	N	Y	N	N
Glue in place	N	Y	N	N
Protect zones	N	Y	N	Y
Decouple caps	N	Y	N	N
Opposite side	Y	Y	N	Y
Add on fly	Y	Y	N	Y
Ratsnest	Y	Y	N	Y
Vector force	Y	N	N	N
Hide paths	N	Y	N	N
Auto pan	Y	N	Y	Y
Zoom range	100:1	64:1	two levels	unlimited
Routing Features				
Autoroute				
Nonstop	Y	Y	Y(2)	Y
Interactive	Y	Y	Y(2)	Y
Pad to pad	Y	Y	Y(2)	N
Mix widths	N	N	N	Y
Lee	Y	Y	Y(2)	N(3)
Heuristic	N	Y	N	N
Rip-up router	Y	Y	Y(2)	Y
Number of copper layers	16	30	2	59
Ortho off	N	N	N	N
Curved tracks	N	N	N	Y
Auto vias	Y	Y	N	Y
Ground planes	Y	Y	Y	Y
Protect zones	Y	Y	N	Y
Neck down	Y	Y	N	N
Editing Features General				
Find/search	Y	N	N	Y
Jump	Y	N	N	Y
Change colors	Y	Y	Y	Y
Macros	Y	N	N	Y Place
Move	Y	Y	N	Y
Rotate	Y	Y	Y	Y
Opposite side	Y	Y	N	Y
Card outline	Y	Y	Y	Y
Change package	N	Y	N	N
Change pad	Y	Y	N	N
Pair swapping	N	Y	N	N
Gate swapping	N	Y	N	N

	OrCAD/PCB II	PADS-PCB	PCBoards	ProCAD
Pin swapping	N	Y	N	N Block
Move	Y	Y	N	Y
Delete	Y	Y	Y	Y
Copy	Y	Y	Y	Y
Save	Y	Y	N	Y Route
Delete track	Y	Y	Y	Y
Change width	Y	Y	Y	Y
Chg via size	Y	Y	N	Y
Via optimize	Y	Y	N	Y
Rubberband	Y	Y	N	Y
Density map	N	Y	N	N
Histogram	N	Y	N	N
Design Rule Check				
Counts number of vias	Y	Y	Y	N
Counts number of nets	Y	Y	Y	N
Reports unconnected nets	Y	Y	N	Y
Reports broken tracks	N	Y	N	Y
Reports air-gap violations	Y	Y	N	Y
Output Formats				
HPGL plotter	Y	Y	Y	Y
HI plotter	Y	Y	Y	Y
PostScript	Y	N	N	Y
Dot-matrix printer	Y	Y	Y	Y
Gerber	Y	Y	N	Y
N/C Drill	N	Y	N	Y
DXF	N	Y	N	Y

	Protel-Autotrax	Protel-Easytrax	Schema-PCB Layout	Tango Tango PCB Plus
Standard Features				
Price	$1,295	$450	$975	$895
Integrated schematic capture	software	N	N	NN
Hardware protection key	Y	N	Y	Y
Copy protected	N	N	N	N
Supports EMS RAM	4MB	4MB	8MB	32MB
Hard disk required	N	N	Y	Y
Help screen	N	N	N	Y
Maximum board size (inches)	32 x 32	32 x 32	32 x 32	32 x 32
Netlist Features				
Front annotate	N	N	Y	N
Back annotate	N	N	Y	N
Bill of materials	Y	Y	Y	Y
Parts mask	Y	Y	Y	Y
Solder mask	Y	Y	Y	Y
Silkscreen masks	Y	Y	Y	Y
Component Placement				
Auto place	Y	N	Y	N
Glue in place	Y	N	Y	N
Protect zones	Y	N	Y	Y

	Protel-Autotrax	Protel-Easytrax	Schema-PCB Layout	Tango Tango PCB Plus
Decouple caps	N	N	Y	N
Opposite side	Y	N	Y	Y
Add on fly	N	N	Y	N
Ratsnest	Y	Y	Y	Y
Vector force	N	N	N	Y
Hide paths	Y	Y	Y	Y
Auto pan	Y	Y	N	N
Zoom range	100:1	100:1	64:1	unlimited
Routing Features				
Autoroute				
Nonstop	Y	N	Y	Y
Interactive	N	N	Y	Y
Pad to pad	Y	Y	Y	N
Mix widths	N	N	N	N
Lee	Y	N	Y	Y
Heuristic	N	N	Y	Y
Rip-up router	Y	Y	Y	Y
Number of copper layers	8	8	30	8
Ortho off	Y	Y	N	Y
Curved tracks	Y	Y	N	N
Auto vias	Y	Y	Y	Y
Ground planes	Y	Y	Y	Y
Protect zones	Y	Y	Y	Y
Neck down	N	N	Y	N
Editing Features General				
Find/search	Y	Y	N	Y
Jump	Y	Y	N	N
Change colors	Y	Y	Y	Y
Macros	Y	Y	N	N Place
Move	Y	Y	Y	Y
Rotate	Y	Y	Y	Y
Opposite side	Y	Y	Y	Y
Card outline	Y	Y	Y	Y
Change package	Y	Y	Y	N
Change pad	Y	Y	Y	Y
Pair swapping	N	N	N	N
Gate swapping	N	N	Y	N
Pin swapping	N	N	Y	N Block
Move	Y	Y	Y	Y
Delete	Y	Y	Y	Y
Copy	Y	Y	Y	Y
Save	Y	Y	Y	Y Route
Delete track	Y	Y	Y	Y
Change width	Y	Y	Y	Y
Chg via size	Y	Y	Y	Y
Via optimize	Y	Y	Y	Y
Rubberband	Y	Y	Y	Y
Density map	N	N	Y	N
Histogram	N	N	Y	N

	Protel-Autotrax	Protel-Easytrax	Schema-PCB Layout	Tango Tango PCB Plus
Design Rule Check				
Counts number of vias	N	N	Y	N
Counts number of nets	N	N	Y	N
Reports unconnected nets	Y	N	Y	Y
Reports broken tracks	Y	N	Y	Y
Reports air-gap violations	Y	N	Y	Y
Output Formats				
HPGL plotter	Y	Y	Y	Y
HI plotter	Y	Y	Y	Y
PostScript	Y	Y	Y	Y
Dot-matrix printer	Y	Y	Y	Y
Gerber	Y	Y	Y	Y
N/C Drill	Y	Y	Y	Y
DXF	N	N	Y	Y

(1) Using optional router
(2) Using <PCRoute.>
(3) Interacative CAD defines it as a probe router (see Chapter 13).

Circuit Design Rating Summary

Schematic Capture Program Ratings

	Drawing	Editing	Library	Netlist Support	Ease of Use
CapFast	excellent	excellent	excellent	excellent	fair
DC/CAD	good	excellent	good	poor	good
EE Designer III	excellent	excellent	good	poor	fair
FutureNet-5	excellent	excellent	excellent	excellent	good
HiWIRE II	fair	fair	fair	fair	fair
OrCAD/STD III	good	good	excellent	excellent	good
ProCAD	excellent	excellent	good	excellent	fair
Protel-Schematic	excellent	excellent	excellent	poor	excellent
Schema III	good	good	good	excellent	good
Schema-Quik	good	good	fair	excellent	good
SuperCAD	good	good	fair	fair	good
Supersketch Plus	excellent	excellent	good	none	excellent
Tango-Schematic	good	excellent	excellent	poor	good

PCB Layout Program Ratings

	Place	Route	Library	Netlist Support	Ease of Use
DC/CAD	good	good	good	good	good
EE Designer III	excellent	excellent	poor	fair	fair
HiWIRE II	poor	good	fair	good	fair
Labcenter PCB II	good	poor	excellent	none	excellent
OrCAD/PCB II	fair	good	excellent	excellent	good
PADS/PCB	excellent	excellent	good	excellent	fair
PC Boards	poor	fair	poor	poor	fair
ProCAD	fair	excellent	fair	poor	fair
Protel-Autotrax	excellent	excellent	good	fair	excellent
Protel-Easytrax	poor	fair	good	none	excellent
Schema-PCB Layout	excellent	excellent	good	good	fair
Tango-PCB Plus	poor	excellent	excellent	poor	good

Glossary

Schematic Capture Terms

Arc: A curved line form of any size that is from 1 to 360 degrees in length. A circle is a 360-degree arc.

Bill of Materials: A file generated for a schematic by a post-processing program. For each component in the design, the file lists the component type, the number of components of each type, and the labels assigned to each component. The file is usually given the extension .BOM.

Block: A section of the workspace that can be marked and then manipulated as a single entity.

Border: The border around the worksheet — a holdback from the hand-drawn drafting days when it took many pages to describe a design, each with its own defined border and page number.

Bus: A line that is used to show multiple parallel lines on the schematic such as address and data lines. Buses are only visual aids for reading a schematic, as opposed to a schematic filled with finely-spaced wires, and are sometimes ignored during post processing.

Child: In an hierarchical design, the level below the current level, getting farther from the root, is the child.

Circle: A 360-degree arc.

Component label: A unique name for a component that provides a way to identify individual components on the schematic. Components are labeled with an initial string of one or more letters followed by a number. Generally, integrated circuits (ICs) begin with the letter U (for example, U1), resistors begin with R (R2), capacitors begin with C (C3), connectors begin with J (J1), switches begin with SW (SW3), and crystals begin with XTAL (XTAL2). Power components, such as Vcc and GND, do not have labels. The component label is also called a reference designator.

Component type: The name of a component as specified in the component library. Examples of component types include 74LS00, LM555, 2764, DB9, CAP, DIODE, and GND.

Component value: Specific electrical information about a discrete component. The possible values that can be assigned discrete components include resistance (for resistors), capacitance (for capacitors), frequency (for crystals), power rating, voltage rating, and tolerance.

Dashed Lines: Lines that are used to show relationships on the schematic. For example, you can use dashed lines to show that a particular pair of relay contacts is controlled by a single relay coil, or that a number of switch poles are operated in parallel. Dashed lines are only visual aids for reading a schematic and are ignored during post processing.

DRC: See **Design Rule Check.**

Dotted lines: Lines that are used to show relationships on the schematic. Most often, dotted lines are used to show a more tenuous relationship between objects than dashed lines. Dotted lines are only visual aids for reading a schematic and are ignored during post processing.

Flat design: A schematic design whose file structure is simple, with every physical item on the board being represented by a corresponding item found on a single sheet.

Free port: In an hierarchical design, a port which corresponds to and is the child of a **module port**.

Global port: In an hierarchical design, a port which is free-floating and extends through the hierarchy from the current level down.

Hierarchical design: A schematic design whose file structure may be complex, where multiple instances of the some physical item on the board or IC may be represented by a single representation of the item on the sheets. A hierarchy may contain an unlimited number of levels and an unlimited number of sheets. However, most schematic programs have a practical level and sheet limit, usually of 1000 or less.

Hidden pins: Component pins that are defined in the component definition in a library, but are not displayed on the worksheet. The post-processing program uses the pin name of a hidden pin to connect it with other hidden pins of the same name, to the net of the same name. Generally, hidden pins are connections to the power rails (usually Vcc and GND).

Junction: A dot on the schematic that connects intersecting wires. The junction distinguishes a single net from two independent nets which cross over each other.

Labeled connections: A means to connect a component pin to an net by drawing a wire from the pin header and placing a label on the wire.

Level: In as hierarchical design, a collection of section which all have a common **parent**.

Menu: A box of commands that can be displayed and removed from screen.

Module: In and hierarchical design, a representation of a **child** section.

Module port: In an hierarchical design, a port attached to a module.

Multi-part component: An IC package that contains two or more elements (such as op amps or logic gates).

Net: A path of connectivity. A net consists of two or more pins that are electrically connected.

Netlist: A file of component connections generated from a schematic. The file lists net names and the pins which are a part of each net in the design. The file also list attribute information about all the components on the schematic. This file is usually given the default extension .NET.

Net name: A text string that is placed on a wire to indicate the name of the net that contains the pins connected to that wire.

Non-orthogonal: A mode for drawing lines, wires or buses. In non-orthogoal mode, you can draw lines, wires or buses at any angle.

Orthogonal: A mode for drawing lines, wires or buses. In orthogonal mode, you can only draw lines, wires or buses that run vertically and horizontally perpendicular to each other; no other angle is permitted.

Parent: In an hierarchical design, the level above the current level, getting closer to the root.

Part: A basic circuit element, such as a NAND gate or a flip-flop, which when combined with other similar or dissimilar parts, forms a component.

Physical connection: A line on a schematic that shows an electrical path.

Pin: A pin is used to describe both a tangible item (like the physical pin or lead of a component) and an intangible concept (like the node in the net). Component pins are distinguished by pin designators, which are simple pin names.

Port: In a hierarchical design, the connection between a child and a parent level are maintained by module ports, free ports ar global ports.

Power components: Examples of power components include Vcc, Vss, +5, -5, +12, -12, and GND.

Reference designator: A unique name for a component that provides a way to identify individual components on the schematic. Typically, components are labeled with an initial string of one more letters followed by a number.

For example, integrated circuits (ICs) begin with the letter U (U1), resistors begin with R (R2), capacitors begin with C (C3), connectors begin with J (J1), switches begin with SW (SW3), and crystals begin with XTAL (XTAL).

Root: In an hierarchical design, the highest level of the hierarchy.

Schematic: A drawing or set of drawings that shows an electrical circuit design. The individual drawings that make up the entire schematic are called sheet files.

Section: In an hierarchical design, a collection of one or more schematic sheets within a level.

Solid line: Solid lines are used to show relationships on the schematic. For example, you can use dashed lines to show that a particular pair of relay contacts is controlled by a particular relay coil or that a number of switch poles are operated in parallel. Lines are only visual aids for reading a schematic and are ignored during post processing.

Title block: A rectangular outline on the worksheet that contains information about that drawing, like the schematic's name, identification number, revision number, sheet size, date, filename, and designer. The title blocks always appears in the lower right corner of the worksheet.

PCB Terms

Block: A section of the workspace that can be marked and then manipulated as a single entity.

Click: The click applies to positioning the cursor on a part of the display and pressing the **Mouse button.**

Component: A set of primitives (such as pads, tracks, and text strings) that are treated as a single entity. A component, which represent an electronic part, is identified on the PC-board by a unique reference designator.

Component library: A special file that contains the patterns for a number of components.

Density: The density of a PC-board is a measure of how tightly packed the components are on the board. By convention, density is given as the number of square inches per equivalent IC (EIC); the lower these number the denser the board. This term is not to be confused with track density, which refers to how close together tracks on the PC-board can be.

Design rules: Every design disciple has a set of rules, known as design rules. For printed circuit boards, the basic design rules specify the minimum clearance that may be between items in different nets (clearance rules), and the way in which the item are to be connected are to be connected (electrical rules). Other rules are the track width required to carry a given current, the maximum length of clock lines, the termination requirements for signals with fast rise and fall times, among others.

Design Rule Check: A design rule check (DRC) ensures that the design rules have been followed. This may be done manually by inspecting the artwork visually, or using automatic testing equipment (ATE).

DRC: See **Design Rule Check.**

Electrical check: This is the process of checking the printed circuit board to make sure that all connections match the net list.

Equivalent IC (EIC): The number of equivalent ICs is a standard method for determining the number of component pins on the PC-board and dividing by 16 (which is considered to be the typical number of pins on a descrete IC).

Excellon N/C: A drill file format that robotic drilling machines use to locate and drill properly sized holes in the printed circuit board.

Force vector: A line emanating from a component on the board whose length and direction are the weighted average of all connections to that component. Force vectors provide a visual cue as to where components should be moved to minimize total connection length and thereby help to achieve optimum component placement.

Free pad: A free pad is any pad that not part of a library component. As a result, it can only be referred to by its position on the board instead of a specific component name, like U1.

Gerber: A file format used by virtually all PCB service bureaus to translate PCB layout files into finishe printed circuit boards using photoplotting methods.

Grid: A grid is an array of dots or lines that provide a means to precisely position the cursor when editing a printed circuit board.

Ground plane: See **plane**.

Highlight: A technique used to emphasize an item on the display. On color systems, a unique color is generally used to highlight an item. On monochrome systems, the highlighted item is generally shown in bold.

IC: An integrated circuit. Integrated circuits combine a number of electrical elements on one chip.

Net: A net is a set of two or more pins that are electrically connected.

Netlist: A file generated from a schematic. The file lists net names and the pins which are a part of each net in the design. The file also lists information about at the components on the schematic, including component type, label and package type.

Node: A node is a pin in a net. In the netlist, a node is described by a component reference designator and a pin designator, separated by a delimiter, as in the example U1-16. A node should only appear once in a net list; otherwise, it implies that the node is more than one net, and those nets should be combined into one net.

Pad: Although pins and pads are commonly used interchangeably, it is important to distinguish between the two. Whereas a pin is used to describe both a tangible item (the physical pin or lead of a component) and as an intangible concept like the node is a net, a pad always refers to a physical shape on the printed circuit board. Pads are available in a variety of shapes and sizes, to suit the component that will be soldered to it. See also **surface pad** and **through-hole pad**.

Pan: To move across the workspace by displaying successive section of a printed circuit board on the screen.

Pin: Although pads and pins are used interchangeably, it is important to distinguish between the two. Whereas a pad always refers to a physical shape on the PC-board, a pin is used to describe both as a tangible item (the physical pin or lead of a component) and an intangible concept (the node in a net). Component pins are distinguished by their pin designators, which are simply pin names like U1-16.

Plane: Most of the connections on a printed circuit board are made with tracks. However, boards will often have planes of copper as their inner layers that are used as a power or ground connection. If a component lead must pass through a plane, the plane has a hole big enough to clear the lead. Component leads can connect to a plane with either a direct connection or a **thermal relief**. Typically, a digital printed circuit board will have two dedicated planes: power and ground. Many analog boards have more than two planes.

Ratsnest: An appropriately named display of all connections on the board made on a point-to-point (unrouted) basis. Like force vectors, the ratnest help the designer improve component placement by visually indicating areas of connection density on the board.

Reference designator: A unique name for a component that provides a way to identify individual components on the printed circuit board. Typically, components are labeled with an initial string of one or more letters followed by a number. For example: integrated circuits (ICs) begin with the letter U (U1), resistors begin with R (R2), capacitors begin with C (C3), connectors begin with J (J1), switches begin with SW (SW3), and crystals begin with XTAL (XTAL2).

Restricted area: There may be places on the board where you do not want tracks to be routed. A restricted or keep out area is an area that restricts track routing within that area.

Scale: Lets the user enlarge or reduce a printed circuit board screen or size the layout when printing or plotting via increments. For example, 2X.

Schematic: A drawing or set of drawings that show an electrical circuit design. The individual drawings that make up the entire schematic are called "worksheets".

Surface pad: Surface pads only appear on the top and Bottom layers of the printed circuit board. These pads are used for edge connector fingers and surface-mount components.

Thermal relief: A means of connecting a component pin to the power or ground plane electrically but not thermally. The thermal relief is a small island of copper around a component lead, isolated from the power or ground plane by an annular gap. The plane is then connected to the island by two or four narrow bridges of copper (termed "spokes"). These spokes provide the electrical connection and the gap provides the thermal isolation.

Through-hole pad: Unlike a **surface pad**, a through-hole pad belongs to all layers and accommodates components with leads that pass through the PC-board.

Trace: A trace is a node-to-node connection which consists of one or more tracks.

Track: A track is a copper line on the printed circuit board. A track has a start point, an end point, a width, and a layer. Tracks can be used to make electrical connection on the board. A complete node-to-node connection is called a trace. This trace is made up of one or more tracks.

Via: A via is a hole through the PC-board that does not have a component lead through it. The via is used to make a connection from a track on one layer to a track on another layer.

Workspace: The workspace is the greater part of the screen. When you load a PCB file, the PCB layout is displayed on the workspace.

Zoom: Used to magnify or contact the workspace of a section of the workspace. Zooming in on the workspace allows you to view a section in greater detail. Zooming out on the workspace allows you to view a wider area.

Index

Trademarks

CapFast is a trademark of Phase Three Logic, Inc., Beaverton, OR.

DC/CAD is a trademark of Design Computation, Inc., Farmingdale, NJ.

EE Designer is a trademark of Visionics Corp., Santa Clara, CA.

FutureNet is a trademark of Data I/O Corp., Redmond, WA.

Hiwire is a trademark of Wintek Corp., Lafayette, IN

IMS is a trademark of Voltec, Inc., Anaheim, CA.

ISIS is a trademark of Labcenter Electronics, R4 Systems, Inc., West Hill, Ontario, Canada.

IsSpice is a trademark of Intusoft, San Pedro, CA.

OrCAD is a trademark of OrCAD Systems, Hillsboro, OR

PADS-PCB is a trademark of CAD Software, Inc., Littleton, MA.

PCBoards is a trademark of PCBoards, Birhmingham, AL

PCRoute is a trademark of RGH Software Design, Baton Rouge, LA.

ProCAD is a trademark of Interactive CAD Systems, Santa Clara, CA.

Protel is a trademark of Protel Technology Inc., San Jose, CA.

PSpice is a trademark of MicroSim Corp., Irvine, CA.

Schema is a trademark of Omation, Richardson, TX.

SuperCAD is a trademark of Mental Automation, Inc. Bellevue, WA.

SUSIE is a trademark of ALDEC, Newbury Park, CA.

Tango is a trademark of ACCEL Technologies, Inc. San Diego, CA.